Journal of Language, Identity, and Education

Volume 3, Number 4, 2004

Special Issue: (Re)constructing Gender in a New Voice
Guest Editor: Juliet Langman

This journal is abstracted or indexed in *EBSCOhost Products; Linguistics and Language Behavior Abstracts; Communication Abstracts; ERIC Current Index to Journals in Education; Cabell's Directories; PsycINFO/Psychological Abstracts.*

First published 2004 by Lawrence Erlbaum Associates, Inc.

Published 2019 by Routledge
2 Park Square, Milton Park, Abingdon, Oxon OX14 4RN
52 Vanderbilt Avenue, New York, NY 10017

Routledge is an imprint of the Taylor & Francis Group, an informa business

ISBN: 978-0-8058-9514-8 (pbk)
ISSN 1534-8458

JOURNAL OF LANGUAGE, IDENTITY, AND EDUCATION, *3*(4), 235–243

(Re)constructing Gender in a New Voice: An Introduction

Juliet Langman
University of Texas, San Antonio

"You look like a woman and talk like a man." This comment, made by a trusted colleague at a Hungarian university where I had been teaching for over two years, led to my initial reflections on the relationship between language learning and gender identity. As I began to achieve high proficiency in Hungarian, my fifth language, in the mid 1990s, I was puzzled by increasing discomfort in my interactions with colleagues at that university. What was puzzling to me—my increasing linguistic skills were leading to less rather than more understanding and cooperation—was for him a simple, psychological fact. For my largely monolingual and monocultural audience, my physical and linguistic identities were in conflict. Then it was my turn to reflect: were my pragmatic and sociolinguistic abilities not keeping pace with my phonological and morphosyntactic advances? I was, after all, able to pass for a native speaker in brief encounters in a variety of social settings. Or was the matter far more complicated than that? Was I engaged in creating a "third culture" (Kramsch, 1993) resisting the stereotypical gender identity offered in my new culture as "too feminine" because it was, crucially, in conflict with my professional identity? Was conflict perhaps the ideal positioning for me given that "passing" would not allow me to "remain myself in a new voice?" Alternatively, perhaps I was still only a peripheral participant in my new academic community of practice, in spite of the fact that I had become a legitimate and central participant in my primary social community of practice, which revolved around the world of traditional dance. Reflections on these questions led me to construct a life narrative that accounted for the conflicts between my diverse identity practices. My own narrative tied the contradictions into a coherent whole that allowed me to be both a participant observer in one community as well as a professor lecturing on issues of language and identity in another. The coherence

Requests for reprints should be sent to Juliet Langman, University of Texas at San Antonio, Division of Bicultural–Bilingual Studies, 6900 N. Loop 1604 West, San Antonio, TX 78249–0653. E-mail: jlangman@utsa.edu

derived from my ability to see participation in my second culture as bounded by diverse communities. In order to maintain my own voice in these different communities I varied my responses to the ideologies of gender and language I encountered.

The exploration of the circumstances surrounding the production of individual narratives about second language learning is the central focus of this special issue. This issue draws together articles that examine the relationship between gender identity and second language learning from a variety of perspectives, all of which share a basic grounding in sociocultural theories of learning and poststructural theories of language. That is, all of the articles in this issue focus on the interplay between gender identity, discourse practices in local communities of practice, and the ideological forces that mediate between them.

In the last several years, new perspectives on the process of second language acquisition have emerged that challenge the dominant view of second language learning as essentially an individual internal cognitive process. In sociocultural theories of learning, the focus of research is on the individual as a sociocultural being whose discourse practices influence the nature of the second language learning experience, as well as the ultimate outcomes of such learning (Lantolf, 1996, 2000). From this perspective, motivation has been reconceptualized as investment (Norton, 2000; Norton Peirce, 1995), and language learning as socially situated discourse.

Pavlenko and Lantolf (2000) employ the metaphor of self-translation to articulate their theory of second language learning. They outline a trajectory for the (re)construction of self in a new language. Following an initial phase of loss that Pavlenko and Lantolf equate with *second language learning,* there comes the phase of recovery and (re)construction that they equate with *second language becoming.* Self-translation occurs as one seeks practices that allow for the translation of chosen identities into new practices in new languages, as well as the redefinition of practices to fit new or changing conceptions of identity.

A second metaphor for learning, participation, is drawn from sociocultural theory and more specifically from the community of practice model (Lave & Wenger, 1991; Wenger, 1998). Norton and Toohey (2002) summarize the goal of future research drawing on a participation model as follows:

> to develop understandings of learners as both socially constructed and constrained but also as embodied, semiotic and emotional persons who identify themselves, resist identifications, and act on their social worlds. Learners' investments in learning languages, the ways in which their identities affect their participation in second language activities, and their access to participation in the activities of their communities, must all be matters of consideration in future research. (p. 123)

One way to examine second language learning as participation, which can be conceived of as a dynamic negotiation between the individual and the community, is through narrative:

While communities construct narratives that often become conventionalized and integrated into the community fabric as cultural models that provide cohesion for the community, individuals also construct their own personal narratives, based on the conventionalized models, which allow them to make their own lives cohesive; that is, to understand what they are and where they are headed. (Pavlenko & Lantolf, 2000, p. 160)

A second influence on recent research on second language learning comes from feminist and poststructural theories that have traditionally addressed the differential ways in which women and men use language. These theories have been applied to second language learning in examining differential access to resources and identities in new discourse settings (Ehrlich, 1997; Norton, 2000; Pavlenko, 2002; Pavlenko, Blackledge, Piller, & Teutsch-Dwyer, 2001). Pavlenko (2002) suggests

the two-way relationship between language and identity recognizes that languages serve to produce, reproduce, transform and perform identities, and that linguistic, gender, racial, ethnic and class identities, in turn, affect the access to linguistic resources and interactional opportunities, and ultimately, L2 learning outcomes. (p. 298)

The community of practice approach, introduced to the field of language and gender by Eckert and McConnell-Ginet (1999), focuses on gender as an identity practice, and in turn, shifts the focus away from an examination of linguistic behavior as a reflection of social categories such as gender. This perspective has the advantage of foregrounding intergroup variation, that is, it neither assumes gender difference nor that there is necessarily a gendered explanation for behavior (Bergvall, 1999; Bucholtz, 1999). Studies of second language learning, grounded in local discourse practices, have the potential to avoid essentialist assumptions on the role that gender plays in any given discursive practice (McElhinny, 2003).

Nonetheless, in recent work in the area of gender and second language learning at least one generalization does emerge, namely, "the persistence of social and cultural practices across various communities ... that adversely effect women's access to interactions in a target language or the nature of such interactions" (Ehrlich, 1997, p. 436). New questions for second language learning outlined by Pavlenko (2001a) seek to examine how gender may play a role: "What are the ways of indexing and performing gender in different speech communities and can they be appropriated by adults? What are the implications of assuming new gendered subjectivities in the target culture?" (p. 139). These questions lead to broader ones that focus on ideologies of language and gender and their effect on identity practices:

What—possibly conflicting—gender ideologies and discourses are at work in a particular society? How are they reflected in language ideologies and linguistic practices? ... Are gender ideologies currently in the process of change and, if so, how is this change reflected in linguistic practices? (p. 142)

The articles in this special issue address these questions by examining the potential effects of gender identity on the process of second language learning in the case of bilingual and immigrant women in a variety of settings. They focus on two areas crucial to our understanding of the process of second language learning: (a) narrative as data and (b) the tension between essentialism and the social constructivist view of identity as multiple and dynamic. In addition, they present a variety of analytic lenses applied to narrative.

NARRATIVES AS DATA: PERFORMANCE OF IDENTITY

Narratives give insight into the experience of second language learning from the learners' perspective. Through narratives we can gather data useful to understanding differences among speakers in terms of their learning process and ultimate attainment that cannot be accounted for in other methods of studying language learning. Narratives allow for a nuanced understanding of how individuals perceive access to resources for learning language as well as for constructing new identities within new communities of practice. In these narratives, issues of investment, positionality, and emotionality are accessible to the researcher.

The authors examine narratives as part of larger ethnographic studies of second language learning. Although the researcher as interlocutor represents a specific setting for narrative, it nonetheless allows for an examination of the multiple and often conflicting identity practices that represent ways in which individuals articulate and practice their second language becoming.

Narratives also represent an area in which gender differences are apparent, and through which, especially when including oral narratives, the voices of the powerless and the silenced can be heard. Pavlenko (2001b) outlines the way in which language learning memoirs constitute a gendered genre:

> Female narrators drew on discourses of gender significantly more than the male narrators. In particular, they continuously reflect upon the gendering of the second language socialization, while we have yet to read stories where male narrators ponder upon the implications of their gendered position as husbands, fathers, sons, communicators, or providers in a new culture. (p. 236)

ESSENTIALISM AND MULTIPLE IDENTITIES

Social constructivist and poststructural theorists urge us to move away from essentialism in our examination of the role gender plays in the process of second language learning (Bucholtz, 2003; Cameron, 1998; McElhinny, 2003; Pavlenko & Piller, 2001). Indeed, gender is not always a variable with significance in understanding the sociocultural being, and the task of the researcher is to determine whether and how gender is a factor. Similar criticisms of the concept of native speakerness challenge its appropriateness as a measure for determining language proficiency and dividing the world into native and nonnative speakers. Native speakerness, too, is an ideological construct that favors the status of native speakers over nonnative speakers, and often conflates other concepts such as race and ethnicity within it (Brutt-Griffler & Samimy, 2001). Piller (2002) redefines the concept of passing, not from the perspective of the native speaker norm but of the language users themselves. In this perspective, passing is a practice that is contextually bound. More recently Bucholtz and Hall (2003) have introduced the concept of adequation, which they distinguish from passing in that adequation invests the individual with greater agency as she makes discourse choices designed to both "highlight similarity with a referenced group and simultaneously to obscure any differences from it" (Skapoulli, p. 253). Once the concept of nativespeakerness disappears as a construct, the distinction of bilinguals and second language learners is also no longer valid.

Yet, in spite of social constructivist researchers' views of the world as socially constructed, the narratives presented here demonstrate how individual agents often see themselves as living in a world starkly defined by clear and essentialized ideologies of gender and language that they must respond to. The authors here employ the concepts of passing (Piller, 2002), adequating (Bucholtz & Hall, 2003), or resisting and not participating within a community of practice framework (Canagarajah, 1999; Norton, 2001; Siegal, 1994) as ways of exploring the relationship between the learners' sometimes essentialized narrative and the sociocultural view of "L2 users as agents whose multiple identities are dynamic and flexible" (Pavlenko, 2001b, p. 319).

DIVERSE APPROACHES TO THE STUDY OF GENDERED NARRATIVES

As Lantolf (1996) argues, to develop a broader understanding of second language learning, we need to "let all the flowers bloom." Represented in the articles here are four different approaches to the analysis of language learning narratives.

Skapoulli employs the tactics of intersubjectivity and more specifically the concept of adequation (Bucholtz & Hall, 2003) to examine the ideologies of gender that mediate discourse practices in the case of an Arabic-speaking adolescent immigrant girl in Cyprus. Nadia, who has lived in Cyprus since age four, has the ability to "pass" as a native speaker of Cypriot as well as standard Greek, yet makes strategic decisions on how and when she will do so. Skapoulli brings the agency of the individual to the forefront and examines the stance of subject through the concept of adequation (Bucholtz & Hall, 2003), which she argues provides a better explanation for strategic decisions than the concept of passing.

Skapoulli's analysis also highlights the notion that cultural space is often essentialized for individuals. Hence, for Nadia, her two cultures represent polar opposites, and she as an individual is skilled at negotiating or adequating in each. Skapoulli argues that Nadia makes a choice to be "in the middle" between these two polar ends and in this way is able "to exercise a critique of both gender discourses and challenge their principles" (p. 256).

Vitanova blends a Bakhtinian approach with a feminist perspective to focus on the ever-shifting dialogic relations between the self and the Other. In this way she highlights the fact that gender identities are relational and liminal. Vitanova takes her analysis further by drawing on Bahktin's notion of emotional-volitional tone which helps to explain the unique experience of individual learners. She argues that such emotions are discursive phenomena that constitute social action, rather than reflections of cognitive states.

Vitanova applies this approach to an examination of gender transformations in the face of second language learning among recent immigrant couples from Ukraine and Russia in the United States. She examines couples' differential adjustments in their gender positionings over the course of two years. Vitanova presents four couples, all highly educated intellectuals in their home countries, who arrived in the United States with limited knowledge of English. She presents a two-tiered analysis, focusing on the role of gender in relation to the native speaker of English, on the one hand, and to the partner in the particular heterosexual couple, on the other. Her analysis reveals that "gender is enacted on two planes of discourse: sensitivity to social positionings, with a particular focus on emotions, and linguistic expertise within the couples" (p. 262). She finds that although the women positioned themselves and were acknowledged as the linguistic experts in the couple, they tended to experience negative emotions of fear, shame, and nervousness in their interactions with native speakers, whereas the men "aligned themselves with the 'legitimate' members of their L2 community" (p. 267). Vitanova concludes that although immigrant women routinely occupy doubly-marginalized positions as women and as second language learners, their discourses of emotion do not necessarily represent a sign of weakness as their discourse also encodes resistance and potential future action. In articulating their feelings, "they are engaging in a complex rhetorical relationship with a specific audience and at the same time are expressing a socioideological position" (p. 274).

Warriner employs a feminist poststructuralist approach to investigate the relationship between ideologies of language and gender that mediate social and economic relations in the case of three Sudanese refugee women. Warriner situates her article in the discourse on language learning and refugee resettlement, a discourse which has paid insufficient attention to the experiences of women and girls. Warriner's analysis rests on work-related narratives which challenge societal-level discourses on recent immigrants' goals, motivations, and abilities. She examines the ideology of English-only that claims that learning English will ensure success in U.S. society, and outlines the structural impediments that make such an ideology a myth.

Her analysis of interview data from a larger ethnographic study draws on Holland and Lave (2001) to focus on "the relationship between the construction of individual identity and the influence of historical structures by focusing on locally contentious practices" (p. 282). As with Vitanova, Warriner incorporates the concept of dialogism, here arguing that her subjects' talk "becomes dialogic by making use of historically influenced structures and processes (such as those expressed in certain ideologies of language, diversity, and immigration) while also influencing and structuring the course of history in a locally contingent manner" (p. 283).

Unique to Warriner's study is the focus not only on gender practices but also on gendered work practices. She highlights the fact that "learning English in and of itself is not enough to enable the women to establish economic self-sufficiency within the U.S. context, in large part due to the gendered aspects of their individual identities" (p. 282). In her analysis, Warriner focuses on the "particulars and complexities of any one person's narrative accounts as well as the specifics of 'strategic situational negotiations'" (p. 283) in opposition to essentializing narratives seeking to characterize, for example, refugees in the United States. Hence, Ayak positions herself as an "assertive problem-solving wife and mother." In contrast, Alouette, a single mother of eight, represents herself as a woman constrained by circumstances. Her narrative represents that of the working poor as well as the internalization of the English-only view. Finally, Moría brings a professional identity as an accountant from Sudan and positions herself as someone who wants to work, or be in school, and not be at home. She "ventriloquates complaints often made about current government policies" (p. 290) that are particularly hard on the working poor.

Finally, Menard-Warwick explores the concept of investment (Norton, 2000; Norton Peirce, 1995) and presents two opposing paths to the decision to study English. She examines the relationship between ideologies of gender and gendered practices. She demonstrates how two women raised in a context with a similar ideology of gender engage in distinct practices with respect to studying English after immigration to the United States. For Trini, learning English is tied to her personal independence and represents progress. Trini's life story is one of resistance to male influences and the implicit ideology of gender that requires obedience on the part of women. She resists, first by going to the United States and later by attending English classes. In contrast, Camila bounded by the same ideology of gender, articulates a narrative of fear and obedience. She initially resists and finally reluctantly acquiesces to learning English in order to do what is right for her family, based on her husband's conception of family and gendered expectations. Coming back to Norton's concept of investment, Menard-Warwick suggests that in spite of these different narratives, both women's decisions to study English derived from their primary identities as wives and mothers. In each case, their decision to study English was an investment in their children and the maintenance of a stable home.

CONCLUSION

Two limitations of this special issue should be noted. The first is the continued focus on English, and more specifically, the United States, in studies of adult immigrants although Skapoulli as well as other recent work does focus on other countries and on non-English speakers (Bilaniuk, 2003; Pujolar, 2001). The second limitation is the focus on female immigrants in all but one article (Vitanova). This prevents us from examining the similarities in experience of all adult immigrants and perpetuates the focus on difference and/or women in gender studies (Ehrlich, 1997; McElhinny, 2003).

Despite these limitations, these articles do present a range of approaches to questions of the role of gender identity in a set of quite distinct local contexts. Cameron (1998) argues that we need theories of gender capable of changing people's customary ways of thinking. To this we might add that we need theories of gender that explain the processes by which individual learners develop and practice new linguistic forms for the purpose of either maintaining or changing customary ways of thinking in new contexts. An examination of the tensions between new and old ways of expressing identity that shifts in code and context evoke will allow for continued theorizing on the nature of gender identity and its role in multiple language learning and use.

ACKNOWLEDGMENTS

I am grateful to Thomas Ricento, Kelly Graham, and the authors of the pieces here. Our insightful conversations in oral and written form have made this process both thought provoking and fun.

REFERENCES

Bergvall, V. (1999). Toward a comprehensive theory of language and gender. *Language in Society, 28,* 273–293.

Bilaniuk, L. (2003). Gender, language attitudes, and language status in Ukraine. *Language in Society, 32,* 47–78.

Brutt-Griffler, J., & Samimy, K. (2001). Transcending the nativeness paradigm. *World Englishes, 20,* 99–106.

Bucholtz, M. (1999). "Why be normal?": Language and identity practices in a community of nerd girls. *Language in Society, 28,* 203–223.

Bucholtz, M. (2003). Sociolinguistic nostalgia and the authentication of identity. *Journal of Sociolinguistics, 7,* 398–415.

Bucholtz, M., & Hall, K. (2003). Language and identity. In A. Duranti (Ed.), *A companion to linguistic anthropology* (pp. 369–394). Oxford, England: Blackwell.

Cameron, D. (1998). Gender, language and discourse: A review essay. *Signs, 23,* 945–975.

Canagarajah, A. S. (1999). *Resisting linguistic imperialism in English teaching.* Oxford, England: Oxford University Press.

Eckert, P., & McConnell-Ginet, S. (1999). New generalizations and explanations in language and gender research. *Language in Society, 28,* 185–201.

Ehrlich, S. (1997). Gender as social practice: Implications for second language acquisition. *Studies in Second Language Acquisition, 19,* 421–446.

Holland, D., & Lave, J. (Eds.). (2001). *History in person: Enduring struggles, contentious practices, intimate identities.* Santa Fe, NM: School of American Research Advanced Seminar Studies.

Kramsch, C. (1993). *Context and culture in language teaching.* Oxford, England: Oxford University Press.

Lantolf, J. (1996). Review Article: SLA theory building: "Letting all the flowers bloom!" *Language Learning, 46,* 713–749.

Lantolf, J. (2000). Introducing sociocultural theory. In J. Lantolf (Ed.), *Sociocultural theory and second language learning* (pp. 1–26). Oxford, England: Oxford University Press.

Lave, J., & Wenger, É. (1991). *Situated learning: Legitimate peripheral participation.* Cambridge, England: Cambridge University Press.

McElhinny, B. (2003). Theorizing gender in sociolinguistics and linguistic anthropology. In J. Holmes & M. Meyerhoff (Eds.), *The handbook of language and gender* (pp. 21–42). Malden, MA: Blackwell.

Norton, B. (2000). *Identity and language learning: Gender, ethnicity and educational change.* New York: Longman.

Norton, B. (2001). Non-participation, imagined communities, and the language classroom. In M. Breen, (Ed.), *Learner contributions to language learning: New directions in research* (pp. 159–171). Harlow, England: Pearson Education.

Norton, B., & Toohey, K. (2002). Identity and language learning. In R. B. Kaplan (Ed.), *The Oxford handbook of applied linguistics* (pp. 115–123). Oxford, England: Oxford University Press.

Norton Peirce, B. (1995). Social identity, investment and language learning. *TESOL Quarterly, 29,* 9–31.

Pavlenko, A. (2001a). Bilingualism, gender, and ideology. *International Journal of Bilingualism, 5,* 117–151.

Pavlenko, A. (2001b). Language learning memoirs as a gendered genre. *Applied Linguistics, 22,* 213–240.

Pavlenko, A. (2002). Poststructuralist approaches to the study of social factors in second language learning and use. In V. Cook (Ed.), *Portraits of the L2 user* (pp. 277–302). Clevedon, England: Multilingual Matters.

Pavlenko, A., Blackledge, A., Piller, I., & Teutsch-Dwyer, M. (Eds.). (2001). *Multilingualism, second language learning, and gender.* Berlin: Mouton.

Pavlenko, A., & Lantolf, J. (2000). Second language learning as participation and the (re) construction of selves. In J. Lantolf (Ed.), *Sociocultural theory and second language learning* (pp. 155–177). Oxford, England: Oxford University Press.

Pavlenko, A., & Piller, I. (2001). New directions in the study of multilingualism, second language learning, and gender. In A. Pavlenko, A. Blackledge, I. Piller, & M. Teutsch-Dwyer (Eds.), *Multilingualism, second language learning, and gender* (pp. 17–52). Berlin: Mouton.

Piller, I. (2002). Passing for a native speaker: Identity and success in second language learning. *Journal of Sociolinguistics, 6,* 179–206.

Pujolar, J. (2001). *Gender, heteroglossia and power: A sociolinguistic study of youth culture.* New York: Mouton.

Siegal, M. (1994). Second-language learning, identity and resistance: White women studying Japanese in Japan. In M. Bucholtz, A. Liang, L. Sutton, & C. Hines (Eds.), *Cultural performances: Proceedings of the third Berkeley women and language conference* (pp. 642–650). Berkeley, CA: Berkeley Women and Language Group.

Wenger, E. (1998). *Communities of practice: Learning, meaning and identity.* Cambridge, England: Cambridge University Press.

JOURNAL OF LANGUAGE, IDENTITY, AND EDUCATION, 3(4), 245–260

Gender Codes at Odds and the Linguistic Construction of a Hybrid Identity

Elena Skapoulli

University of California, Santa Barbara

This study traces the ways in which an Arabic-speaking immigrant girl in Cyprus positions herself towards the different gender ideologies that she encounters at home and in school. Through Nadia's linguistic choices and discursive strategies we come to see how competing gender codes, meeting at the crossroads of geographic, linguistic, and cultural transition, lead to the emergence of a hybrid cultural identity. The data also reveal that "passing" is not always the goal of speakers who enter a new culture; instead, social categories such as gender mediate between the ability and the willingness to pass. These findings suggest that L2 users, through their daily linguistic and cultural practices across and within discourse sites, become agents of multiple, dynamic, and flexible identities. The study demonstrates the particular implications of second language learning and use and gender ideologies in the process of identity construction and highlights the complexities of identity work in today's multilingual and multitextual social settings.

Key words: identity, gender, hybridity, second language learning, passing

The need for new directions in conceptualizing gender has been supported by a number of scholars in the field of language and gender (Bucholtz, 1999a; Cameron, 1998; Piller & Pavlenko, 2001). They all stress the importance of retheorizing gender beyond hegemonic and essentialist perspectives that have often prevailed in the field. The problematization of commonly held assumptions about gender becomes even more imperative within the context of multilingual and multicultural societies, as in such contexts the interaction of language and gender is further complicated by historical and ideological processes that set the parameters for the expression of self. As these social spaces provide a variety of languages and contexts with diverse expecta-

Requests for reprints should be sent to Elena Skapoulli, Gevirtz Graduate School of Education, 2206 Phelps Hall, University of California, Santa Barbara, CA 93106–3100. E-mail: eskapuli@education.ucsb.edu

tions of gender behavior, they often lead to transformations of women's and men's understanding of gender and positioning of self in discursive interactions.

One of the questions recently raised in the field of language and gender is how gender is implicated in the appropriation of new discursive practices in the process of learning a second language (L2) (Pavlenko, Blackledge, Piller, & Teutsch-Dwyer, 2001). Pavlenko (2001a) argues that women L2 users may respond to new gender discourses by assimilating or resisting existing ideologies of gender in either language, a practice which at times leads to the creation of new hybrid identities that deconstruct the links of language, gender, and ideology.

This study, focusing on the linguistic practices of Nadia, an Egyptian teenage girl in Cyprus, extends this argument by showing that advanced L2 female speakers may deploy their competency to negotiate multiple identities according to their own agendas about gender practices. Through Nadia's sophisticated use of Greek, in which she expresses and enacts her selective detachment from the gendered practices of her native-Cypriot peers, we come to see the affirmation of a hybrid cultural identity which subverts the assumption that an individual's goal in a new culture is to "pass" as native. These observations contribute to a deeper social understanding of language contact phenomena and shed light on linguistic settings beyond the range of the existing scholarship on language and gender and second language acquisition (SLA), which have traditionally focused primarily on Anglo-American contexts.

"PASSING"

One of the concepts that has been extensively dealt with in the context of diverse societies is the notion of "passing." Sociolinguistic research on identity performance has explored the concept of passing with respect to gender or sexual identity (Barrett, 1999; Hall, 1995), ethnic identity (Bucholtz, 1995; Wieder & Pratt, 1990), and recently, with respect to the social identity of advanced L2 learners who can pass as native speakers (Piller, 2002a). In these studies passing is conceptualized as the ability to be perceived as a native member of a particular community, which results in the accomplishment of particular social goals of the speakers, such as their social integration in contexts where ethnicity becomes critical.

Passing is always implicitly associated with the notion of authenticity and the subsequent assumption that certain linguistic features are emblematic to particular social, ethnic, or gender groups. Although the concept of authenticity has been widely challenged (Bucholtz, 2003; Pratt, 1987; Woolard, 1998) and strongly contested with the discussion on hybrid identities (Jaffe, 2000; Piller, 2002b; Rampton, 1995), the correlation of passing with some kind of a "prototype norm," that is, with "language produced in authentic contexts by authentic speakers" (Bucholtz, 2003, p. 398) still resonates in the relevant literature. In addition, it is usually assumed that speakers enjoy certain benefits by passing as "authentic" members of a particular community and therefore would more likely prefer to pass. This assumption leaves unexamined the practices of speakers who, despite their high competency, articulate a "nonauthentic" identity and yet access symbolic benefits associated with passing, such as social integration. This article suggests that passing is not always the goal of speakers who enter a new culture; instead, social categories such as gender, mediate between the ability and the willingness to pass. The act of "adequation" (Bucholtz & Hall, 2003) can better explain the nuanced positions that an individual takes on in a hybrid situation like the one discussed here. Adequation denotes the relation that establishes sufficient sameness between individuals or groups and allows for a transient partial identification rather than a recategorization of the individual from one social group to another. Whereas passing depends in many ways upon the *audience's* perceptions and evaluations, adequation underscores *speakers'* assertion of similarity.

LANGUAGE, GENDER, AND SECOND LANGUAGE LEARNING AND USE

The discursive practice of identities has become one of the central foci of much of the recent body of sociolinguistic research that examines speakers' representations of self and negotiation of identities through language. In particular, the growing field of language and gender has extensively applied discursive approaches to identity and contextual analyses of the relationship between language and gender. The introduction of the model "Communities of Practice" developed by Lave and Wenger (1991) and refined in the realm of gender identity by Eckert and McConnell-Ginet (1992) epitomizes this approach. Further, it suggests that the issues should be addressed through examination of the everyday social and linguistic practices of local communities that provide particular sets of gender practices to their members. The question now becomes how language *effects* gender (Bucholtz, 1999a), and within the conceptualization of this study, how access to and use of different languages shapes gender performance.

Despite the proliferation of research on language and gender there has been little attention to the role of second language learning and use in the process of gender identity construction, and particularly, in social contexts outside the Anglophone world. While today multilingual settings constitute the normative social space in which the performance of identity takes place, the history of monolingualism and anglocentrism, which has dominated the field of language and gender for the past 30 years, has not nurtured a fertile ground for the examination of gender in the process of immigration or in other kinds of intercultural contact.

Piller and Pavlenko (2001) argue that ideologies of language and gender, inherently ingrained in language, provide new gendered discursive practices and understandings of gender identities to speakers of a new language. The notions of gender-appropriate linguistic behavior, gendered practices, and the meanings of femininities and masculinities implicated in "doing" and performing gender (Butler, 1990; West & Zimmerman, 1987) are not shared across cultures. Learning a second language may often entail a modification of one's gender performance in order to ensure validation in the new culture. And particularly within the context of immigration, where certain identities may be rendered inaudible, participation in new communities of practice may lead to the development of new identities, or subject positions (Pavlenko, in press).

Extending this line of reasoning, this study suggests that the process of second language learning (SLL) should be involved in the dialogue about the interaction between gender and language. The basic assumption of the present study is that women and men who experience geographic and cultural transition, encounter new ideologies of gender embedded in the new language and culture, which require them to rethink and revise their understanding and discursive performance of gender.

One of the key components of this argument refers to the ways in which gender structures new interactional opportunities for female and male L2 users. Individuals who develop and practice new linguistic forms automatically create a new voice to express self in social interaction. As a result, L2 learning multiplies possibilities of gendered self-expression and provides opportunities for agency and choice in the performance of self (Pavlenko, 2001b). However, the context of cultural transition is often accompanied by a clash of ideologies of language and gender, or power asymmetries among speakers that complicate this picture. For example, despite the potentiality of choice in the performance of gender, the power differential in various contexts may impose certain constraints to the range of identities that an individual is able to take on. In addition, the journey from a traditional culture to a society of late modernity, or vice versa, may make the links between gender identity and language choice and use more problematic and more apparent. This observation suggests that the relationship between SLL and gender should be examined through ethnographic approaches that take into account the complexities of specific contexts and the individuals' understandings of them. In particular, researchers should focus on learners' *use* of a second language, as individuals may learn a new language and use it selectively according to their goals in social interaction. In addition, apart from the incorporation of SLL in the discussion, the field of language and gender must also extend its focus by including research inquiry on noncustomary contexts and populations, such as social spaces at the periphery of Europe and the United States with non-English speakers.

METHOD

This article draws on a corpus of interview data collected for a larger ethnographic study that focused on the ways in which five immigrant teenage girls from Russia,

Ukraine, Georgia, and Egypt describe and interpret their experience in the schools of Cyprus (Skapoulli, 2002). Cyprus, an island of approximately 700,000 Greek- and Turkish-speaking inhabitants in the eastern corner of the Mediterranean, is currently in the process of multicultural transformation with the reception of large numbers of economic immigrants from the Middle East, Southeast Asia, and Eastern Europe and with the prospect of incorporation in the European Union in May 2004. The larger study sought to explore, among others, immigrant girls' social relations with the Cypriot peer group and their strategies of adaptation in the school social context. The initial analysis of the data prompted further exploration of the practices of Nadia, a 16-year-old Egyptian girl, who, in contrast to the other participants, articulates an uneasy relationship with the range of gender practices of the mainstream local youth.

Nadia arrived in Cyprus at the age of 4, following her family's journey for economic advancement. She comes from the Coptic Christian minority of Egypt which numbers several hundred people in Cyprus. The Copts, a strong religious entity, who pride themselves on being fervent defendants of the Christian faith, have their own churches in the island's major cities where they conduct weekly masses and social events. Nadia's native language is Arabic, which is spoken at home and in community gatherings, but she does not define herself as Arab—which in her view is synonymous with Muslim. During the course of the study Nadia attended a prestigious high school in Nicosia, the capital of Cyprus, and all her peers and friends there were native Cypriots.

Despite Nadia's long presence in Cyprus (12 years), her high fluency in Greek (L2), and her native-like accent, she does not embrace the idea of passing as native, as she associates that with particular gender practices which she renounces. In fact, all the consultants in the larger study, and other students who were asked about this issue informally, characterized gender performance as the determining factor for one's integration in the mainstream peer group. Angela, one of the immigrant girls who participated in the research project, describes the prerequisites of membership in the local youth group as a "common way of thinking," which she explains as "dressing in a modern way, going out to nightclubs and cinemas, and buying expensive clothes and shoes." Although one cannot talk about a "global" peer culture in the Cypriot school context, the study's consultants consider the cultural practices of the *mainstream* student group, and particularly those involving the performance of gender, as the seminal features of "the" peer culture.

These remarks support Eckert's (2001) argument about the importance of the "heterosexual market" in the youth world. Eckert suggests that as young teenagers approach adolescence, they find themselves in a market place where their value is largely determined by their visibility to and contact with the other sex. This context propels youth to produce themselves as commodities in a heterosexual market which entails adopting particular styles of gender behavior that are most "marketable." According to Eckert, one of the resources they employ to do so is language.

Particular forms of linguistic behavior maximize one's value in the peer group and ultimately give rise to a peer-based linguistic market. Thus the school context becomes a site of competition for visibility and popularity which sometimes involves a great degree of creativity.

Such creativity is manifest in the way Nadia uses the different cultural resources that she encounters at home and at school. In what follows, I discuss Nadia's self-appropriations of the linguistic and social practices that inform her struggle to forge a gender identity. The paper addresses the following questions

1. How does Nadia make use of the diverse gender discourses that she encounters in the various communities that she navigates?
2. What is the relation between her gendered linguistic practices and the processes of second language learning and use?

The analytical framework of the study draws on Eckert and McConnell-Ginet's (1992) paradigm of "communities of practice," where gender is conceptualized as the product of the interaction between language and other symbolic systems, such as dress, body adornment, ways of talking, ways of moving, places for hanging out, and so on. Gender, within this framework, is constructed through the social practices that people employ in the multiple communities in which they are members. This line of thinking makes possible a more comprehensive examination of gender identity in the context of cultural and linguistic contact, as it allows for the consideration of the multiple spaces in and means by which negotiation of gender takes place.

METHODOLOGICAL CONSIDERATIONS

The study's interest about the role of the local peer culture and L2 use in gender identity performance has delineated the scope of questions and language used in the interviews. All interviews were conducted in Greek and dealt with issues of gender practice, primarily within the school context. As I claim though that individuals portray a gendered self through language and that each language affords speakers with particular gendering discourses, accordingly, Greek is expected to influence Nadia's expression of gender differently than Arabic. Therefore, the data only *partially* reveal the spectrum of Nadia's discursive performance of gender. The study's purpose, however, is to illuminate the process by which immigrant girls negotiate new discourse conventions and gender practices central for their social integration in the new community—one of these processes being the act of passing (or not) as native in Cyprus—therefore, Nadia's performance of gender in Arabic is beyond the study's analytical focus.

Apart from the language used in the interviews, the power differential was another factor that has been taken into account both throughout the interviews and in my subsequent communication with Nadia. Being asked by a Greek-Cypriot woman/researcher to define her relationship with the local gender practices, Nadia might have confronted a dilemma on how to negotiate a possible disaffiliation from the Cypriot code of femininity. Partly, this limitation was redressed with the intimacy Nadia and I gradually developed and with the reflexive character of the study. During the 18 months of the study I conducted four in-depth interviews with Nadia, of approximately 45 minutes each time. I have also maintained regular telephone communication with her since then, which offered me the chance to consult with her about my interpretations of the data.

HYBRIDITY AS "ADEQUATION" WITH COMPETING GENDER CODES

Bucholtz and Hall (2003) propose the term "adequation" to describe speakers' pursuit of sufficient similarity and proximity with a referenced group, though not necessarily complete sameness or identification with it. The interview data reveal that Nadia's struggle to form a gender identity involves adequation with competing gender codes. Her participation in different communities of practice—the local Western culture in the school setting and her Eastern home culture—places the challenge of accommodating to gender codes at odds. Nadia crystallizes this conflict as centered on issues "pertinent to adolescence." On the one hand, her peers invite her into the world of Western youth culture that the global media portray, which entails dating, clubbing, wearing makeup and sexy clothes, going to the cinema, and buying popular culture products. On the other hand, her parents admonish her to be quiet, reserved, and modest, and follow the Coptic principles according to which a woman, symbolized by the metaphor of a blooming flower, should remain "untouched" (*aneggichti*) and "immaculate" (*pentakathari*) up to her marriage.

Despite their internal complexity and heterogeneity, these social worlds are governed by certain gendering systems with the relevant consequences on individual behavior. Being simultaneously a member of a conservative religious community and of a liberal peer culture inevitably creates tensions and complications for a teenage girl's development of gender identity. The interview data reveal that Nadia responds to this challenge by embracing aspects of both home and peer culture gender practices, a strategy that results in the emergence of a hybrid cultural identity. She materializes this stance in the use of specific social and linguistic practices that she employs to display her affiliation with or disaffiliation from each culture.

The following analysis highlights both the linguistic and nonlinguistic constitution of gender and shows that the process of gender identity formation centrally involves cultural beliefs regarding language use and interaction. Language, apart from being a crucial resource for the performance of identity, is connected to an entire network of practices, ideologies, and social relations. As Ochs argues (1992), linguistic practices that index gender do not occur in a vacuum; they are filtered through social values and belief systems. This discussion demonstrates that the legitimization of specific forms of talk with their particular symbolizations of gender function in tandem with cultural ideologies embedded in broader sociocultural contexts.

HYBRIDITY AND SPEECH STYLE

Nadia's interactional choices and gender practices portray the emergence of a hybrid cultural identity as a response to the challenge of having to balance gender codes at odds. In the journey of constituting a gendered self, Nadia adopts certain social practices of Western youth culture. She makes use of beauty commodities such as mascara, hair gel, and hair removal products; she buys mainstream teenage magazines, and she uses virtual chat rooms on the Internet frequented by youth from all over the world. Her behavioral patterns are equally influenced by the gender expectations of the Coptic religion. As she describes in the interviews, she avoids wearing miniskirts, low-cut shirts and bikinis, reading or discussing magazines that portray male models, and going out to nightclubs. Both cultural codes therefore inform her understanding of gender and infuse their values in her ways of expressing a gendered self.

The social practices described here are coupled with linguistic strategies that support her effort to balance incongruent codes of femininity. Nadia reports how her linguistic behavior varies according to context and she also expresses awareness of this variation with metalinguistic comments about the different speech styles. For example, she asserts that she may use profanity with her colleagues when she is at work but avoids doing so with people from the Coptic community. Through adequation, Nadia brings to salience the commonalities with her interlocutors in each social setting to "fit in." Thus in linguistic terms Nadia may pass in either social group by manipulating her speech style accordingly. The following example illustrates how Nadia describes her contextual adaptation.

1 I try to/ to change my character wher- depending on the place I am.
2 for example, on Sunday when I go to church, I am not like this at all, o.k.?
3 I try to talk and laugh but I try to watch my mouth too/ not to act goofy all the time.
4 Whereas here in school we say silly things all the time/ we talk very very openly.
(See Transcription Conventions on page 260.)

Nadia here distinguishes the contexts of church and school as the two polar institutions that drive her ability to maneuver within different sets of practices. Church as the ultimate site of instantiation of the Coptic culture and school as the space of contact with the local peer group represent the major arenas where she feels the conflict of contradictory beliefs about being a woman. A reserved and modest speech behavior allows Nadia to meet the expectations of her parents and other members of the Coptic community (line 3) while an extrovert and carefree speech style helps her to blend in with the peer group (line 4). The data reveal that as gender ideologies are embedded in language use and situated in social relations, they make possible different patterns of femininity through the use of different speech styles. Nadia's ability to manipulate diverse "gender selves" in diverse contexts agrees with earlier research findings (Pavlenko, 1998, 2001a; Pavlenko & Lantolf, 2000) that delineate individuals' multiple identities as subject to transformation with change in communities of practice.

Nadia's strategic contextual adaptation also confirms Bucholtz & Hall's (2003) assumption that adequation is a *motivated* social achievement that speakers employ to with a referenced group and simultaneously to obscure any differences from it. And particularly for an immigrant girl who participates in multiple communities of practice—in some of which certain identities may not be easily enacted or accepted—there seems to be an emergent need for flexibility and adaptability. Evidently, the social landscape is a primary determinant for one's experience of gender and when this landscape is formed with hills and valleys that demand variable accommodations, its navigation may lead to multiple positionings along the road, or otherwise, to multiple identities in practice.

HYBRIDITY AND L2 COMPETENCY

The previous discussion highlights the gendering discourse that different registers afford Nadia in the process of gender performance. This section extends this line of reasoning by drawing attention to Nadia's use of certain linguistic features emblematic to the local youth culture. The data reveal that Nadia's high competency in Greek provides her with the discursive tools, in terms of vocabulary, prosody and style, to display her relationship with the different gender discourses and manipulate language according to her goals. One of the ways she does that is through the selective use of standard Greek or the Cypriot dialect of Greek in her reference to the different gender discourses.

In Cyprus people use standard Greek in formal contexts such as the school classroom and at times when they want to mark an encounter as formal. At all other times they use the local dialect of Greek, which is also the primary register of young people. Standard Greek is associated with propriety, social distance, and sophistication, whereas the local dialect is associated with casualness and intimacy. In her interactions with me, Nadia only shifts her register to the standard at times when she addresses issues related to Coptic principles, as in the following example, where she responds to my question "What does your culture expect from a woman?"

14 *prepei i kopela na meinei aneggichti, na min tin aggizei kanenas [...]* a girl
    should remain untouched, no one should touch her [...]
19 *na einai i kopela pentakathari [...]* a girl should be immaculate [...]
22 *dioti tin vlepoun kati poly tryfero, kati aneggichto ktl* because they view her
    as something very tender, something untouched, etc.

This statement, and other references to the Coptic belief system, is entirely articulated in standard Greek, in contrast to most of the interview data where the Cypriot
dialect is used. Nadia invokes the formal register to describe the refined code of femininity of the traditional Coptic culture (line 14 *"aneggichti"*; line 19
"pentakathari"; line 22 *"tryfero"*). In contrast, she uses a colloquial dialect to describe her relationship with the gender practices of youth, for example, in describing
her reaction to the surprise party that her friends organized on her birthday: *"pou tin
pollin charan mou arkepsa na tous filo oullous"* (I felt so much joy that I started kiss-
ing all of them). As my question above invites her to become the representative of traditional Coptic culture, she invokes a formal Greek register that can better portray a
serious spokesperson of her community, while when she talks about her immersion
in the peer culture she adopts the language that directly reflects youth values.

   The incorporation of English expressions in her speech produces further evidence of Nadia's participation in the youth linguistic market, a practice that is common among young people in Cyprus. As the local argot is saturated with features
of Western pop culture, primarily from the music and movie industries, English
loanwords provide symbolic resources for youth to claim membership in a global
youth culture. Nadia demonstrates her familiarity with this practice when she dis-
cusses topics related to youth life. For example, when she confided in me that she
was romantically interested in a boy she said *"en ixero inta pou sou to lalo tora
alla,* **anyway,** *areskei mou"* (I don't know how come I'm telling you about this now
but, **anyway,** I like him). English loanwords are used in the context of youth enter-
tainment and social or romantic relationships. Nadia mentions an occasion when
her classmates organized a **surprise party** for her on her birthday; she says that
she doesn't see any **chance** that the boy she likes would approach her, and she reports that she enjoys it when a boy feels **relaxed** and confides his secrets to her.
These examples not only indicate Nadia's access to the local argot of the young but
they also reveal the linguistic resources that the peer culture provides speakers for
talking about issues related to youth cultural practices, and particularly, to the
practice of romance. In so far as language is part of the packaging of a product for
the local heterosexual market, as Eckert (2001) argues, Nadia draws on linguistic
resources, such as English loan words, colloquial expressions, and slang, to display her membership in this market and express a romantic self.

   Nadia's strategic use of the two varieties of Greek demonstrates her ability to
represent herself in different ways: as a blithe teenager or as a member of a conservative community. These observations point to the role of L2 competency for the

negotiation of multiple identities in practice. Not only does Nadia know the vocabulary specific to multiple registers but she also knows when and how to use it. Not only does she talk about romance but she also uses the appropriate style to talk about it. Not only does she talk about the Coptic gender expectations but she also uses the appropriate language to do so. The highly interactional register of the youth culture is represented through Nadia's casual expressions and bodily cues while the more conservative Coptic culture is represented with a formal lexicon, standard Greek, and a serious quality in her voice and posture.

In this case, therefore, the term "second language learning *and use*" is more appropriate than "second language acquisition" since a speaker may learn a new language but use it selectively and strategically according to context. For Nadia, second language use is inextricably connected to gender identity and directly associated with social context. Nadia's deployment of L2 suggests that for a multilingual speaker, gender identity can be displayed through multiple ways, *iconic* (Irvine & Gal, 2000) to different gender ideologies. As Irvine and Gal propose, linguistic features are often used to iconically embody social images, and thus further reinforce the ideological link that binds them together. In this case, competing gender codes, reflected in Nadia's varied language use, point to the two polar gender discourses of tradition and modernity. As each culture affords Nadia with different linguistic resources to express her gendered self, a hybrid identity thereby emerges.

"IN THE MIDDLE": DISTINCTION AND BALANCE

Nadia describes this hybridity as living "in the middle." She consistently refers to her stance as a "balanced" position between the two worlds and mitigates any evaluation she makes that might subvert this equilibrium. Often her assertions of affiliation with the peer culture are followed by an immediate statement that expresses her disposition not to be perceived as one of her Cypriot peers. This attitude is restricted to certain gender practices of the local group, and particularly, to issues regarding sex and the body. As the prior examples reveal Nadia *selectively* manifests her membership in the mainstream peer culture and the determining factor for this selectiveness is her personal view of gender performance. In short, despite her comfortable use of the youth register and her strategic use of gender practices, she chooses not to identify with the local peer group—not to "pass" as native Cypriot—an identification that she associates exclusively with the mainstream youth's gender practices. Similarly, she articulates her distinction from the traditional aspects of her home culture. Her response to the question "How would you describe your match with the youth culture?" is representative of this pattern:

23 I myself am trying to comp- to:: not to be too exaggerated
24 but neither be too, u::h distant,

25 like, neither be completely completely in the:: exactly have a boyfriend etc.
26 and like not to be exactly exactly like the-
27 like I told you; in the middle
28 neither very much West
29 nor very much East
30 in the middle

In this excerpt, Nadia's simultaneous affirmation of the middle and negation of the poles operate in multiple levels of discourse. At a referential level, Nadia explicitly expresses her positioning with respect to the two worlds with the repetition of the statement "in the middle" as lines 27 and 30 indicate. Apart from confirming her position, Nadia also provides a clear picture of how she perceives them; apparently, she considers them polar opposites, a perception that is evidenced with the use of the contrastive metaphors of "West" and "East." At the semantic level, in addition to the meanings that she ascribes on the two cultural ideologies (as "too exaggerated" and "too distant"), she also implies their quality using the established indices for each, that is, West portraying a more liberal culture and East representing a rather traditional mentality. Furthermore, matching up the pairs of sentences in lines 23–24, 25–26, and 28–29 that juxtapose the two worlds, one can clearly see that the West (line 28) is associated with the practice of romantic relationships (line 25) and is portrayed as "too exaggerated" (line 23). Given that the West is mostly related to the peer culture, as the previous sections revealed, what Nadia here affirms is that she does not want to be "completely completely" part of it. On the other hand, being wholly identified with the East means being "too distant" (line 24) which she also denies in line 26. By using similar concepts of extremity ("completely" and "exactly") and by pointing to the end of the continuum ("exaggerated" and "distant") Nadia expresses her reluctance to identify wholeheartedly with either. The depiction of the two worlds as extreme poles implies that the middle is the most sensible position, thereby implicitly justifying her choice of "distinction" from the two worlds.

According to Bucholtz and Hall's (2003) tactics of intersubjectivity distinction, the strategy speakers use to underscore difference, often operates in a binary system that allows individuals to create an alternative to either pole. Nadia's choice to be "in the middle" demonstrates this practice, which enables her to exercise a critique of both gender discourses and challenge their principles. For example, she clearly states that she cannot connect with the Coptic community because "their mentality is too conservative," but she also critiques her peers, who are often only interested in someone's "figure" or "beauty" instead of valuing a person's character and personality. Her comments throughout the interviews show her opposition to certain aspects of both cultural ideologies as well as her selective appropriation of others. Nadia's stance suggests that distinction can coexist with adequation as at different times individuals may highlight their similarities with or their differences from particular groups.

Her disaffiliation from the local peer culture is nowhere clearer than on issues regarding the body. Perhaps the most prominent aspect of the peer discourse of femininity in Western youth culture, the body is a constant preoccupation of young teenage girls (Bloustien, 2002; Nichter, 2000). This attitude is in contrast to the Coptic belief system, according to which the body is "very revered" and "very sacred" and should not be uncovered or discussed. When Nadia's classmates talk about the body she prefers to avoid such discussions or to "balance things a little":

33 So in these cases I always like to be silent/
34 If there is something for example that I can say
35 (that)I am sure, I am positive for example, that everybody else will agree
36 not to say anything **extreme** and be perceived a:::s strange
37 o.k. I say it.
38 So I try to balance things a little

Nadia's stance taking on the issue of "talking about the body" portrays the balance that she wishes to maintain by keeping silence. Instead of openly taking a stance towards the peer linguistic practice that would require transgression of the symbolic boundaries set by her community's moral code, Nadia merely avoids any active participation in this practice. The use of the English loanword "**extreme**" in her phrasing, which indicates her participation in the local linguistic market, shows that the distance she maintains from her peers resides in the use of *gendered* linguistic practices (talking about sex and the body) rather than in any other linguistic marker of the youth group. In this particular incident, Nadia chooses to enunciate what is locally sanctioned or to stay silent, instead of articulating direct opposition to the group's standards. Nadia's attempt here is not to "pass" as native but to "be passed by" as invisible, since she does not try to present herself as one of the locals by sharing their views, nor does she want to stand out as different or "strange." Although phenotypically she does not differ considerably from her peers and linguistically she easily blends in with the rest of her peers, Nadia agentively chooses not to pass as an "authentic" member of the local peer group.

These observations support Pavlenko's (2001a) assertion that the context of cultural transition may provide women with enough freedom to be "the kind of women they would like to be" (p. 147). Women and girls who encounter new gender ideologies and alternative versions of womanhood at times may cross boundaries and assimilate to the new communities, yet in other situations, the new subject positions may be seen as unacceptable or incompatible with their own models of femininity. Although I do not claim that immigrant women have open access to any kind of identity they wish to adopt, the case of Nadia demonstrates that (trans)formation of gender may start with the adoption of linguistic identities, which enable women to perform self in multiple ways through the use of multiple registers and speech styles.

CONCLUSION

The study has traced the ways in which Nadia, an Arabic-speaking immigrant girl in Cyprus experiments with diverse gender codes and positions herself with respect to different identities. By adopting aspects of both home and peer culture gender codes, Nadia articulates a hybrid identity, which in turn informs her discursive and social practices in the expression of "balance." This attitude points to the notion of adequation, that best fits the analysis of a hybrid situation as it acknowledges the flexibility for partial rather than complete identification with different groups. The findings of this study suggest that any attempt to theorize gender should consider the reflexive nature of the relationship between language and identity; that is, how linguistic choices and practices construct a gendered self, and conversely, how individuals' perceptions of gender identity shape their language practices.

The study brings to light the complex landscape that characterizes multilingual speakers' identity practices in today's plural societies. Nadia's accounts demonstrate her skillful use of L2 that operates in concert to her goals in the process of self-performance. Despite her high fluency in L2 she selects a partial participation in the local linguistic and heterosexual markets since membership in the peer group is associated with certain gender practices that she renounces. Evidently, passing as native is not part of Nadia's agenda, so long as it entails following a particular code of femininity. To reiterate, the data reveal that the intervention of the social category of gender enables new ways of understanding the relationship between passing and SLA that requires us to disentangle the two phenomena and incorporate alternative interpretations.

The findings here support the premise that immigration and other intercultural contact pose a fundamental challenge to individuals: that of finding ways for constructing self in a new voice. To achieve this, immigrant young women may creatively draw on diverse discourse resources from an enormous reservoir of texts in symbolic systems, institutions, and social interactions that allow them to "pick and choose" instances of identities in their social encounters. The emergence of hybrid identities is here conceptualized as a byproduct of an ongoing dialectic where "women and girls do not unthinkingly consume cultural forms but construct their own meanings and identities in relation to such forms" (Bucholtz, 1999b, p. 349). The study confirms the problematics of identity work in today's multilingual and multitextual social settings and the particular implications of gender cultural ideologies in the process of identity construction. Moreover, the study, by drawing attention to a social and linguistic space outside the spotlight of Western scholarship, extends and connects the fields of SLA and language and gender with a view to furthering a productive dialogue between the two.

ACKNOWLEDGMENTS

I am grateful to Professor Mary Bucholtz for her encouragement and invaluable suggestions that brought this article into being. A special acknowledgment must be made to my advisor, Jenny Cook-Gumperz, for her support throughout the writing process. I also thank Nadia, whose openheartedness (*anoichtosyni tis kardias tis*) provided these new insights in academic scholarship.

REFERENCES

Barrett, R. (1999). Indexing polyphonous identity in the speech of African American drag Queens. In M. Bucholtz, A. C. Liang, & L. A. Sutton (Eds.), *Reinventing identities: The gendered self in discourse* (pp. 313–331). New York: Oxford University Press.

Bloustien, G. (2002). Far from sugar and spice: Teenage girls, embodiment and representation. In B. Baron & H. Kotthoff (Eds.), *Gender in interaction: Perspectives on femininity and masculinity in ethnography and discourse* (pp. 99–136). Amsterdam: Benjamins.

Bucholtz, M. (1995). From mulatta to mestiza: Passing and the linguistic reshaping of ethnic identity. In K. Hall & M. Bucholtz (Eds.), *Gender articulated: Language and the socially constructed self* (pp. 351–373). New York: Routledge.

Bucholtz, M. (1999a). Bad examples: Transgression and progress in language and gender studies. In M. Bucholtz, A. C. Liang, & L. A. Sutton (Eds.), *Reinventing identities: The gendered self in discourse* (pp. 3–25). New York: Oxford University Press.

Bucholtz, M. (1999b). Purchasing power: The gender and class imaginary on the shopping channel. In M. Bucholtz, A. C. Liang, & L. A. Sutton (Eds.), *Reinventing identities: The gendered self in discourse* (pp. 348–365). New York: Oxford University Press.

Bucholtz, M. (2003). Sociolinguistic nostalgia and the authentication of identity. *Journal of Sociolinguistics, 7,* 398–416.

Bucholtz, M., & Hall, K. (2003). Language and identity. In A. Duranti (Ed.), *A companion to linguistic anthropology* (pp. 369–394). Oxford, England: Blackwell.

Butler, J. (1990). *Gender trouble: Feminism and the subversion of identity.* New York: Routledge.

Cameron, D. (Ed.). (1998). *The feminist critique of language: A reader.* New York: Routledge.

Eckert, P. (2001). *Linguistic variation as social practice: The linguistic construction of identity in Belten High.* Malden, MA: Blackwell Publishers.

Eckert, P., & McConnell-Ginet, S. (1992). Think practically and look locally: Language and gender as community-based practice. *Annual Review of Anthropology, 21,* 461–490.

Hall, K. (1995). Lip service on the fantasy lines. In K. Hall & M. Bucholtz (Eds.), *Gender articulated: Language and the socially constructed self* (pp. 183–216). New York: Routledge.

Irvine, J. T., & Gal, S. (2000). Language ideology and linguistic differentiation. In P. V. Kroskrity (Ed.), *Regimes of language: Ideologies, polities, and identities* (pp. 35–84). Santa Fe, NM: School of American Research Press.

Jaffe, A. (2000). Comic performance and the articulation of hybrid identity. *Pragmatics, 10,* 33–39.

Lave, J., & Wenger, E. (1991). *Situated learning: Legitimate peripheral participation.* Cambridge, England: Cambridge University Press.

Nichter, M. (2000). Fat talk: What girls and their parents say about dieting. Cambridge, MA: Harvard University Press.

Ochs, E. (1992). Indexing gender. In A. Duranti & C. Goodwin (Eds.), *Rethinking context: Language as an interactive phenomenon* (pp. 335–358). Cambridge, England: Cambridge University Press.

Pavlenko, A. (1998). Second language learning by adults: Testimonies of bilingual writers. *Issues in Applied Linguistics, 9,* 3–19.

Pavlenko, A. (2001a). "How am I to become a woman in an American vein?": Transformations of gender performance in second language learning. In A. Pavlenko, A. Blackledge, A. Piller, & M. Teutsch-Dwyer (Eds.), *Multilingualism, second language learning, and gender* (pp. 133–174). Berlin: Mouton.

Pavlenko, A. (2001b). "In the world of the tradition, I was unimagined": Negotiation of identities in cross-cultural autobiographies. *International Journal of Bilingualism, 5,* 317–329.

Pavlenko, A. (in press). Poststructuralist approaches to the study of social factors in second language learning and use. In V. Cook (Ed.), *Portraits of the L2 user.* Clevedon, England: Multilingual Matters.

Pavlenko, A., Blackledge, A., Piller, I., & Teutsch-Dwyer, M. (Eds.). (2001). *Multilingualism, second language learning, and gender.* Berlin: Mouton.

Pavlenko, A., & Lantolf, J. (2000). Second language learning as participation and the (re) construction of selves. In J. P. Lantolf (Ed.), *Sociocultural theory and second language learning: Recent advances* (pp. 155–177). New York: Oxford University Press.

Piller, I. (2002a). *Bilingual couples talk: The discursive construction of hybridity.* Amsterdam: Benjamins.

Piller, I. (2002b). Passing for a native speaker: Identity and success in second language learning. *Journal of Sociolinguistics, 6,* 179–206.

Piller, I., & Pavlenko, A. (2001). Introduction: Multilingualism, second language learning, and gender. In A. Pavlenko, A. Blackledge, A. Piller, & M. Teutsch-Dwyer (Eds.), *Multilingualism, second language learning, and gender* (pp. 1–16). Berlin: Mouton.

Pratt, M. L. (1987). Linguistic utopias. In N. Fabb, D. Attridge, A. Durant, & C. MacCabe (Eds.), *The linguistics of writing: Arguments between language and literature* (pp. 48–66). New York: Methuen.

Rampton, B. (1995). Language crossing and the problematization of ethnicity and socialization. *Pragmatics, 5,* 485–513.

Skapoulli, E. (2002). *The schooling experience of immigrant young women in Cyprus.* Unpublished master's thesis, University of California Santa Barbara, Santa Barbara, California.

West, C., & Zimmerman, D. H. (1987). Doing gender. *Gender & Society, 1,* 125–151.

Wieder, D. L., & Pratt, S. (1990). On being a recognizable Indian among Indians. In D. Carbaugh (Ed.), *Cultural communication and intercultural contact* (pp. 45–64). Mahwah, NJ: Lawrence Erlbaum Associates, Inc..

Woolard, K. (1998). Language ideology as a field of inquiry. Introduction to B. B. Schieffelin, K. A. Woolard, & P. V. Kroskrity (Eds.), *Language ideologies: Practice and theory* (pp. 3–47). New York: Oxford University Press.

Transcription Conventions

/ slight fall indicating more is to come

- truncation (e.g., what ti- what time is it?)

::: lengthened segments

[…] text omitted

JOURNAL OF LANGUAGE, IDENTITY, AND EDUCATION, 3(4), 261–277

Gender Enactments in Immigrants' Discursive Practices: Bringing Bakhtin to the Dialogue

Gergana Vitanova
University of Central Florida

Drawing on the narratives of four East European couples, this article offers a discourse-centered analysis of their gendered experiences in the second language (L2). The analysis of the data integrates critical feminist perspectives with a Bakhtinian lens to language and the self. Espousing Bakhtin's concepts of dialogue, answerability, and emotional-volitional tone, the article illustrates how the participants in this study enact their positional identities on two planes of L2 use: relative to native speakers of English and within the couples. It shows the discourses these men and women voice in responding to their everyday realities with a particular focus on discourses of emotions and linguistic expertise. It also stresses that these responses are created on the boundary between the self and the Other. The article suggests that a Bakhtinian approach would allow us to add another dimension to viewing the gendered self: that of the ever-shifting relations between the self and the Other which, uniquely to Bakhtin, are fused with a sense of dialogic responsibility.

Key words: gender, second language, adult immigrants, discursive practices, Bakhtin, emotions

Gender has been problematic for scholars across disciplines as is evident in language research (Eckert & McConnell-Ginet, 2003), in psychology (Gergen & Davis, 1997), and in recent interdisciplinary feminist thought (Hesse-Biber, Gilmartin, & Lydenberg, 1999). Common in all these various approaches is the departure from viewing gender as a universal, essentialist category and the emphasis on gender as socially constructed. For example, Ehrlich (1997), who criticizes the traditional models that have governed sociolinguistic research, proposes that we should investi-

---

Requests for reprints should be sent to Gergana Vitanova, 1446 Sunningdale Way, Orlando, FL 32828. E-mail: gvitanov@mail.ucf.edu

gate gender not as an invariable factor, but as an aspect of the social practices of particular communities. Pavlenko, Blackledge, Piller, and Teutsch-Dwyer (2001) have embraced this understanding in second language acquisition (SLA). They adopt the view suggested by first language researchers (e.g., Bergvall, Bing, & Freed, 1996), that gender is not a uniform construction; rather, its sociolinguistic implications vary from one community to another. At the same time, in reviewing gender in second language (L2) and bilingual research, Piller and Pavlenko (2001) write, "gender continues to be under-theorized and under-researched" (p. 3).

This need to theorize the complex relationship between language and gender has been voiced by many feminist researchers. In an inspiring paper, Cameron (1996) writes, "We need theories of gender and of the language-gender interface, that are not just academic renditions of received wisdom, but are capable of challenging people's customary ways of thinking" (p. 49). Reflecting on the difficulty of arriving at a comprehensive theory of gender and language, Bergvall (1999) suggests that to achieve a comprehensive consideration of gender, we need both macro and microlevel analyses and diverse research perspectives on this intricate factor. This means that to build the theories Cameron calls for we must: (a) study this complex interface in its varied contexts of language use; and (b) consider multiple, interactive perspectives in its analysis. In response to these theoretical and methodological calls, this article integrates critical feminist perspectives and a Bakhtinian (1981, 1984, 1993) lens to view the (gendered) discursive practices of four particular couples.

What role does gender play in the L2 worlds of immigrants? How is gender acted out not only in relation to the Other—the native speaker of English—but also in the discursive practices of a particular heterosexual couple? What discourses do the participants employ in response to their everyday realities, and can we discern any gender patterns in the language of their responses? Drawing on the experiences of eight East European immigrants, this article offers a discourse-centered analysis of their subjectivity. Thus, here I am not so much interested in how participants conceptualize gender as a social category, but in how it emerges in the text of their experiences. Specifically, I illustrate how gender is enacted on two planes of discourse: sensitivity to social positionings, with a particular focus on emotions, and linguistic expertise within the couples. My goal is also to elucidate how Bakhtin's notions of dialogue and emotional-volitional tone can enrich our understanding of these phenomena.

THEORETICAL UNDERPINNINGS

The data analyses that follow are guided by critical perspectives on gender and subjectivity and by the writings of Bakhtin (1981, 1984, 1993). In advancing a practice theory of the self, Holland, Lachicotte, Skinner, and Cain (1998) draw on sociocultural perspectives as articulated by Vygotsky (1962), Leontiev (1978), and Bakhtin. Rather than viewing culture as a holistic entity, Holland et al. offer the metaphor of figured worlds, which entail not only "the coproduction of activities, discourses, performances and artifacts" but also "figures, characters, and types" (p. 51). Figured worlds are narrativized, according to the researchers who use ethnographic interviews as the unit for their analyses. In these worlds, selves are dialogic and identities are always positional:

> Relational identities have to do with behavior as indexical of claims to social relationships with others. They have to do with how one identifies one's position relative to others, mediated through the ways one feels comfortable or constrained, for example, to speak to another, to command another, to enter into the space of another, to touch the possessions of another. (p. 127)

The concept of relational selves is not foreign to the epistemology identified by Holquist (1990) as dialogism. Dialogism, "a form of architectonics" (p. 29), refers to the science of relations that encompasses not only aesthetic categories but also self–other relations.

Butler (1990) urged us to view gender as performance. Much earlier, Bakhtin (1981) offered a framework for the study of individual voices as performances in relation to other voices while underscoring the social nature of all texts. Just as people can position their bodies to perform physical acts, they can also position themselves on discursive planes to enact dialogic relationships. Throughout his writings, Bakhtin stresses the interactive nature of discourse and subjectivity: "The dialogic orientation of discourse is a phenomenon that is, of course, a property of *any* discourse" (p. 279). However, as Morson and Emerson (1990) caution, it would be an oversimplification to assume that dialogue in this framework merely implies an oral exchange between two speakers. Instead, in claiming that dialogic relationships are "an almost universal phenomenon, permeating all human speech and all relationships and manifestations of human life—in general, everything that has meaning and significance" (1984, p. 40), Bakhtin offers a philosophical model of viewing the world. In this model, voices represent particular socioideological positions in particular dialogic relationships that are not fixed or immutable but rather shift as power relations change between speaking subjects. Thus selves in this model are always liminal, always occurring on the boundaries between the self and the Other.

Although dialogue is the most discussed Bakhtinian concept, language researchers rarely address another interrelated notion in the Russian scholar's philosophy: answerability. *Otvestvennost'*, as Bakhtin (1994) termed it in his native tongue, (answerability/responsibility), entails the necessity for selves to answer each other's voices in a discursive event and reflects his view of human agency. Of particular importance to this article is that answerability as Bakhtin understood is infused with a sense of moral responsibility toward the Other and with a particular

emotional-volitional tone (a complex of one's desires, feelings, and ethical evaluation). Emotional-volitional tone—a central category in Bakhtin's (1993) early philosophy—is the force of one's acts: "Everything that I have to do with is given to
me in an emotional-volitional tone, for everything is given to me as a constituent of
the event in which I am participating" (p. 33). These emotional-volitional
axiological orientations that seek a response and address other positions underlie
the nature of answerability.

Bakhtin (1993) argued that emotional-volitional tone is not merely a passive
state. In contrast, it orients action and makes answerability unique to a particular
individual. In analyzing the discourses of emotion employed by the participants of
this study, I build on Bakhtin's concept of an answerable, responsible self whose
acts are imbued with emotional-volitional tones. I also espouse his understanding
of authoring in illustrating the discourses these women and men voiced in responding to their new sociocultural realities and thus authoring their voices. As Holland
et al. (1998) expound, "The meaning that we make of ourselves is, in Bakhtinian
terms 'authoring the self,' and the site at which this authoring occurs is a space defined by the interrelationship of differentiated 'vocal' perspectives on the social
world" (p. 173). What meanings do the participants make of themselves and others
in their narrativized responses to everyday realities?

DATA

The Study

This article focuses on aspects of a larger qualitative study (Vitanova, 2002) conducted over 2 years that explored the gender and agency practices of East European immigrant couples and drew on ethnographic interviews (Holland et al.,
1998; Spradley, 1979). Four couples situated in a midwestern city in the United
States participated in the study. At the beginning of the project, all had been in
the United States for approximately 6 months. Table 1 summarizes their background.

None of the participants was fluent in English; the two older women had
studied some English on their own before their arrival. Although the two youn-
ger couples had received some formal instruction in English as part of their high
school and college education, they admitted that it was highly unsystematic.

The interviews were predominantly conducted in English as I wished to elicit
L2 data for another part of the project. However, the participants frequently
switched to Russian. With few exceptions I strove to interview the participants as
couples. This allowed me to not only elicit narrative data but also to gain insight
into the interactional patterns between the partners.

TABLE 1
Description of Participants

Couples	Age	Home Country	Occupation in Home Country	Occupation in Immigrant Country
Vera	Early 50s	Russia	TV and radio journalist	Initially kitchen manager; later: own catering business
Aleksei			High school teacher	Factory mechanic
Sylvia	Late 40s	Ukraine	Communications engineer	Initially department store assistant; later: clerk at a bank
Boris			Architect	Construction worker
Natalia	Mid 20s	Ukraine	Student in business	Part-time student and server at restaurant
Dmitri			Computer scientist	Part-time student and server; later: student and computer programmer
Lydia	Mid-late 20s	Ukraine	Engineer	Part-time student and computer programmer
Peter			Computer scientist	Computer programmer

The Linguistic Other: Positionings in the L2

Analyzing the autobiographies of successful bilinguals, Pavlenko and Lantolf (2000) delineate the various stages through which the writers transformed themselves. A loss of linguistic identity marks the beginning of this transformation, which could also be seen in the experiences of the participants in this study. The eight participants were highly educated and described themselves as intellectuals. When they arrived in the immigrant country, they all experienced a profound loss of their first language voices and reflected on how the L2 (or the lack of it) positioned them socially. Vera, for example, a precise Russian speaker who became a kitchen manager in her immigrant country said, "I had a very, very interesting job. I liked my job. And I understand that here I cannot work as a journalist because I don't know English good." For the two older couples, in particular, the loss of voice also resulted in an overt loss of their standing as intellectuals, and they became acutely aware of the social implications of (not) knowing English.

The term *Other* has become increasingly popular in postmodern social and cultural studies (Riggins, 1997). In a Bakhtinian fashion, Riggins cautions against the illusory distinction between the self and the Other and suggests that the two are closely intertwined on an ever-changing continuum of discourses. All participants indicated awareness of how native speakers of English positioned them as the Other. However, although all reflected on their positions in the new social milieu, the men and women had varying responses to these discursive positionings. This was particularly evident in relation to the use of English as a second language.

Gender and Sensitivity to Positionings

Peirce (1995; see also Norton, 2000) discussed power relations in SLA from a poststructuralist point of view. In her study, for example, the participant Martina did not feel comfortable speaking English as an immigrant. In the larger field of gender studies, Kaschak (1992) has introduced the concept of sensitivity. Building on postmodernism and feminist research, she argues that for historical reasons women have had to remain more sensitive to their environment and particularly to their relations with others. The female participants in this study consistently exhibited patterns of heightened sensitivity to the L2 context. Not only did they reflect on these positioning practices more frequently but they also attached a powerful emotional significance to them. Vera experienced the language practices of otherness in the work environment as exclusive and humiliating. She overtly claimed that the native speakers treated her as *"vtorim sortum"* (second-hand) because of her lack of linguistic skills in English:

> Do you know / sometimes they stay and they are talking in their native language / and I cannot understand because they talk very very fast / and I don't know about / what they are talking / and they ask me something / I cannot answer them because I don't know about what they talk. And / they are looking / "Mm …" Do you know? *Nu / tyajelo* (It's hard) …[1]

As Vera looks at herself through the eyes of her native English-speaking coworkers, her self-consciousness closely resembles the intense self-consciousness of Dostoevsky's hero analyzed by Bakhtin (1984). Vera, who felt uncomfortable in the L2, contrasted her experience with that of a male relative she knew well, juxtaposing his voice with her own:

> He says he needs to teach them [his American coworkers] Russian. It's interesting. My sister's husband Kostya / he works on the factory … And he begins only / maybe three four months / and he works very nice / and / in two months / he begin to work like a supervisor. He don't know one English word! But / he is the supervisor now. And only … and / no one [else there is] / Russian worker. But / when Kostya come every morning / they cry, "Kostya, *privet*! (Kostya, hi!) (Vera laughs). He teach them. He teach them. And now they know / 15 Russian words. And he said, "Vera, my English is very bad. I don't know English. But / be sure / that / in 5 years / they begin to speak only Russian (Vera laughs). I say, "Kostya, you need study English!" He said me, "I am a supervisor! They need to learn my language."

---

[1]See Legend of Transcription Symbols on page 277.

Although Vera's statement that Kostya did not know a word of English was slightly exaggerated (I once met him at a large social gathering), Kostya's English was severely limited, and Vera's English language skills were superior to his. Moreover, through conversations with other members of this immigrant community, I came to realize that Vera's English was highly regarded by other immigrants of her age. Yet despite her better command of English, the disparity between how she and Kostya position themselves in the L2 comes across strongly in Vera's narrative. Kostya does not demonstrate the discomfort that Vera does but assumes a superior position (albeit in a carnivalesque discourse, to borrow a Bakhtinian term) because he is a manager. Vera, on the other hand, who is also a manager in her own work context, feels that it is she who should address the Other in his or her native tongue.

After almost 2 years in her immigrant country, Natalia admitted that she was not always comfortable with Americans in the L2. When her comment prompted me to ask a more general question about their perceptions of belongingness in the context of the new country, Dmitri replied, "This is very unfair question, who belongs, who doesn't belong. Because nobody belongs in this country. This is country of immigrants." His statement closely parallels one made by Boris on another occasion. In response to a comment by Sylvia in which she shared her fear of communicating in English, Boris claimed that all Americans were historically immigrants and asked, "Why not for me?" Both Dmitri and Boris aligned themselves with the "legitimate" members of their L2 community.

Such gender-related patterns also emerged when the participants described their interactions in the L2. For example, Lydia acknowledged that if she asked a person to repeat something and still did not understand, she would not ask again but would rather say, "OK" and then try to "recall" what the person had said. Responding to Lydia, Peter said, "If I still don't understand, I tell them send me the words or whatever but explain me. ... They must explain me. It's their job." Peter felt confident in his rights of a participant in the L2, and he placed the responsibility for understanding in the realm of the Other. Although his misunderstanding might cause inconvenience, it did not result in the feelings of inadequacy experienced by the four women in the study, who often employed discourses of emotions to describe their interactions in the L2.

Discourses of Emotion

As foreigners move across geographic and linguistic boundaries, they cross emotional borders as well. When the participants arrived in their host country, they found that they were not able to participate fully in the dominant language of the new environment. This triggered in them a set of emotional responses, particularly the female participants. Sylvia frequently spoke of her emotions. Fear, nervousness, and shame dominated our conversations about her language experiences in English (e.g., the word *afraid* appears 59 times in her narratives). In the following

excerpt, Sylvia narrated a grocery shopping experience when she could not remember the English word *prunes*:

> S: I was afraid that the cashier wouldn't understand us.
> G: Did you try to say something?
> S: [confirming].
> G: Did they understand?
> S: She understood. [I was] nervous, nervous. I confused because / I always
>   think that we *storonnyie* [groping for words] that we look like ...
> G: Say it in Russian.
> S: *Smeshnie* (funny) [to the others].

Not knowing the English word for prunes made Sylvia feel ashamed. The
shame originated in her perceiving herself and her husband as the Other, thus
"funny" in the eye of the observer. Shame is an emotion that Sylvia often refers to
in relation to English. On another occasion, for example, as she was describing a
telephone conversation in the L2 and her inability to understand the other speaker,
Sylvia exclaimed, "It's a shame!"

Postmodern feminist Bartky (1996) claims that shame in speech is a mark of powerless discourses. She also says that shame requires an audience. This resonates with
a Bakhtinian understanding of the relationship between the self and the Other where
emotions originate as responses to concrete or generalized others. We need the Other
to contextualize our experiences. It is exactly through another that our emotions are
validated, and Sylvia's examples illustrate the nature of this dialogic.

Bartky (1996) also writes that women are more prone to feeling ashamed not
because shame is gender-specific, but because of their historical social location in
powerlessness. I do not suggest that the men do not experience fear, nervousness,
or shame. These feelings are universal. However, a gender-related pattern emerged
from the data in how men and women took up discourses of emotion. This pattern
was prominent in Sylvia and Boris' interactions, and the following example, where
they reflect on their use of English with native speakers, reveals the disparity in
their perceptions:

> S: Take it easy (speaks of Boris).
> G, S: (Laugh.)
> S: I am afraid ==
> B: I no feel guilty ==
> S: I am afraid all the time ==
> B: I no feel guilty. American people / all American people / was / immigrate.
>   Live a few people / now / English. A few. [...] Why not for me?
> S: (Sighs.)
> B: Why / I / must / be guilty? Why?

S: He hasn't any / complexes. It seems to me / I / *kak skazat'* (how do you say)
/ *neudobstvo* (discomfort). *Ya prichinyau / lyudem / neudobstvo* (I cause
people discomfort).

Even as Sylvia was describing being afraid and guilty, Boris negated her stance
and positioned himself on equal terms with the native speakers of English. Whereas
Sylvia was overwhelmed with guilt because she caused another person discomfort
when she groped for words, Boris affirmed his position of a nonnative speaker in a
land of immigrants. Sylvia's feelings were not, however, individual mental states,
nor did her words of emotion express these individual states. Her feelings of shame
and guilt originated in the dialogical processes with other interlocutors.

It was not only Boris and Sylvia's discursive examples that suggested this pattern.
Vera, for example, often spoke of how she felt as a result of loss of voice. Referring to
a quick exchange between her English-speaking colleagues, she commented that
she felt like a fool. In her narrative discourse she would often refer to feelings both in
English and Russian as she described interactions with others in English: "I can **feel**
it" or "*U menya takoe* **chuvstvo** (I have this feeling)" are just two examples. Her husband, on the other hand, never spoke of feelings associated with the use of the L2.
"No problem" is how Aleksei typically described his interactions in English.

Not unlike Vera, when speaking about her communication with others in English,
Natalia commented, "But sometimes when I'm too busy **oh!** I am getting **nervous**, I
can't say anything." The interjection *Oh!* and the adjective *nervous*, also used by the
other women, are both markers of emotion discourses. When Sylvia had to take an
older woman who could not speak English to the hospital for a cardiovascular test,
she had to function as an interpreter between the patient and the physicians. She was
worried that she would make serious mistakes and stressed several times that she felt
afraid. She also repeated the adjective *nervous* several times: "Nervous, nervous,
nervous because I had responsibility! Great responsibility!" In Sylvia's narratives,
discourses of emotions and responsibility are inextricably intertwined. She feels
nervous about communicating in English exactly because of a heightened sense of
responsibility toward somebody else. Sylvia frequently cited responsibility for understanding the other interlocutor. "I feel responsible for understanding" is how she
explained her interactions in the L2. In general, the men did not introduce discourses
of emotion. Of the four male participants only Peter mentioned that he felt uncomfortable when he had recently arrived in the United States and started his job.

To Bakhtin (1993), however, the emotional-volitional tone is "not a passive psy-
chic reaction. ... This is an answerably conscious movement of consciousness,
which transforms possibility into the actuality of a deed (a deed of thinking, of feeling, of desiring, etc.)" (p. 36). As the active force of answerability, the emotional-volitional tone orients the participants' actions. For example, it causes Vera to claim a
space of resistance as she announces, "No, this is not going to be!" in response to native speakers who make her feel uncomfortable in her narrative. A feeling of anger

prompts Lydia to challenge a telemarketer when he comments on "these Russians who can't speak English" and uses an obscenity. And yet Sylvia's desire to respond to and address the Other in linguistically correct L2 motivates her to study her grammar books and check her dictionary as she travels to work on the bus.

Gender and Discourses of Linguistic Expertise Within the Couples

In analyzing the women's use of emotion discourses, the question naturally arises: Do they feel more vulnerable in the L2 because of their lower L2 skills? Perhaps the men, being more advanced speakers of English, did not experience the negative feelings of nervousness, shame, or guilt. Yet further analysis of discourse data belies such a conjecture. Although it was the women who tended to speak of discomfort in the L2, the men typically deferred to their wives' linguistic expertise in English, a trend that emerged throughout the study. For example, the women would introduce metalinguistic terms more often or provide metalinguistic clarifications for their husbands. Vera, for example, clarified for Aleksei's benefit that *French* is the adjective, but *France* is the noun:

> V: I studied in the university so much about Spain / and about Italy / and about Fra ... France?
> G: Yeah, France.
> A: French.
> V: (to A) No, *French eto yazyk. Strana eto France.* (French is the language. The country is France.)
> A: Mm American peoples / all time I listen / speak French French French. Why?
> V: (to A) *Pravil'no govoryat po frantsuskii / strana Francia. Yazyk / French* (It is correct to say "speak French." The country is France. The language is French) [...] *Da, da. And France sushtestvitel'noe.* (Yes, yes, and France is the noun).

In other cases the men would ask their wives for help with vocabulary or with translation from Russian into English. And yet at other times the women would interrupt their husbands to correct their imperfect L2 grammar.

Vera's L2 linguistic superiority over Aleksei was evident in numerous interactions and was accepted by Aleksei himself. Once, for example, as I was talking with Aleksei and Vera was busy nearby, I asked him a question in English. When Aleksei could not understand, he requested help from Vera: "Vera, help me." Language was Vera's domain in the family. Even when Aleksei did not ask for assistance, Vera was ready to offer it as in the excerpt below:

G: (to A) So tell me about your usual day.
V: (translates for him): *Opishi tvoi ...*
A: Ya ponyal ponyal ponyal (I understood.)

Aleksei's tone indicated that this was a familiar pattern. Vera would frequently correct her husband's grammar as illustrated in the following cases:

A: ... He is / understand to me. No problem. He is repair.
V: (to A) *Pochemy* (why) "is"?
A: Why?
V: He repairs.
A: Yeah he repairs.

Vera did not simply correct his error but asked a direct question of him, and Aleksei accepted her linguistic expertise. Having acknowledged his wife's feedback, he repeated the sentence in its "correct" form. Please note that Vera did not necessarily supply the most appropriate form in this context, which would have been "he repaired [it]." It is important for this analysis, however, that she was undeniably the linguistic authority of the couple and was recognized as such by Aleksei.

Sylvia and Boris's interactions also contained excerpts pointing to Sylvia's linguistic superiority in the family. Boris himself repeatedly stated that she spoke English better than he did. Like Aleksei, Boris often required Sylvia's assistance in supplying the needed vocabulary item as illustrated here:

1 B: My boss / give me / exercise [task] for a job / and / *vsegda* (asks S for the English
2   word)?
3 S: Always.
4 B: And always / *zakonchivaet ...*
5 S: Finish / finishes.

Boris directly asked for help with the word *vsegda*. In line 5 Sylvia interpreted his inability to complete the sentence as another request for help and provided the English word. Simultaneously, she exercised her metalinguistic awareness and corrected herself, adding the correct form of the verb: *finishes*. Like Vera, Sylvia provided corrective feedback to her husband as in the following excerpt:

B: And we was in / art museum.
S: We were.

Like Aleksei, Boris acknowledged his wife's linguistic feedback by repeating the supplied correct verb form. Interesting in this and in other similar examples is the automaticity with which she did this. It appeared to be a natural reflex rather than a conscious effort on her part.

The patterns observed in Vera and Aleksei's and Sylvia and Boris' interactions were characteristic of Lydia and Peter as well. Again it was Lydia who provided the corrective feedback to her husband as demonstrated in the following excerpt:

> P: ... and you know / you feel yourself / it's mm how to say? There's like classes of people / and depend on this class mm depend on these classes / people speaking other language ==
> L: Different.
> P: Different different language. Sorry. ... Different language so ...

Like Vera and Sylvia, Lydia monitored her husband's grammar and vocabulary choices. Peter not only recognized Lydia's linguistic authority but also anticipated it as is clear from the following exchange:

> P: And American is living = American is living / I think majority of them / are living for themselves = for theyselves. And that's it. And people are not so hooked up. And / it's very / I think it's very ... *Shto*? (What?) (looking at Lydia)
> L: *Nichego* (nothing).
> P: *Shto-to nepravil'no*? (something incorrect?)
> L: *Ya nichego* (I didn't say anything).

Peter was accustomed to Lydia's feedback, requested or not. In this case he sensed that something was wrong. He even unsuccessfully attempted to correct the error in the first sentence. His second attempt to self-monitor was superfluous as the original form *themselves* was accurate, and the second time he used the incorrect *theyselves*. Lydia refrained from intervening both times. However, being used to her linguistic feedback, Peter expected her to react and pursued it, repeating his question twice. In various cases Peter, much like Boris and Aleksei, would ask Lydia about the English equivalent of a Russian word.

This section reveals that the female participants in this study were the linguistic experts in the couples. Moreover, they were accepted as such by their respective spouses. And yet when communicating in English with others, it was the women who felt shame or fear of making a mistake. Interestingly, I could identify no instance where the women asked their husbands for language assistance.

DISCUSSION

These data and accompanying analyses show how the participants in this study enacted their positional identities on two planes of L2 discourse: that relative to native speakers of English and that between the couples. On one hand, the women displayed a higher metalinguistic awareness and positioned themselves as the linguistic experts within the family. On the other hand, in relation to native speakers, they tended to perceive themselves in discourses of otherness (e.g., Sylvia's feeling that she looked different and, thus, "funny" to the clerk at the grocery store) exactly because of their imperfect, as they saw them, L2 skills. Although these discursive planes may seem contradictory at first, Bakhtin's notion of dialogic answerability/responsibility reconciles them in a way that links the individual and the social. This sense of dialogic responsibility governed the women's desires (to Bakhtin, an inseparable part of one's consciousness) to be linguistically accurate so that they could adhere to the norms of the Other's speech. Ironically, for these participants, who were already marginalized by virtue of being women as feminist scholars would claim, this also tended to increase the perception of power inequality between them and speakers of the L2.

Building on Bakhtin, sociocultural anthropologists Holland et al. (1998) claim that we continually answer what is directed toward us and interpret what is happening. Studying narratives from ethnographic interviews, the researchers assert that as subjects give voice to their experiences and interpretations, they are authoring themselves and the meaning of experience itself. The authoring spaces that the female participants in this study carved for themselves were tinged with emotional-volitional tones. In recent years discursive psychologists (Potter, 1996) have argued that discussions of emotion should move away from the traditional understanding of categories of emotion as individual feelings and referential expressions. Such modern researchers (see also Edwards, 1999) have criticized conceptual analysis of idealized cognitive models where emotions are treated as mental states or cognitive representations. Rather, discourse analysts suggest that emotions are discursive phenomena and should be studied as talk that performs social action.

Such a view of emotions as discursive constructions dovetails with Bakhtin's understanding of human consciousness where no experience can exist outside language. Language, or rather discourse, has a constitutive role in the formation of the self, the self's thinking, and desires. Furthermore, in this perspective, emotions are a boundary phenomenon, born in the relation between the self and the Other. Analyzing Bakhtin's (1993) *Toward a Philosophy of the Act*, Hicks stresses, "In the act of being in which a person relates to objects and other participants in terms of distinction of worth, rational cognition and emotional-volitional tone are co-occurrent" (1996, p. 107). It became clear from the women's discourses that attitudes and emotions originate in the intertextual ground between the self and the Other. For them, much as for Bakhtin's (1984) hero, attitudes are tied to attitudes toward another and to those of another toward them. Their "consciousness of self is constantly perceived against the background of the other's consciousness ... 'I for myself' against the background of 'I for another.' Thus, the hero's words about himself are structured under the continuous influence of someone else's words about him" (p. 207). The women in this study were continually "eavesdropping" on the voices of others, and these axiological orientations structured their perceptions of self in the L2.

Accounting for the role of emotions in gender studies has been a thorny issue for feminist researchers, as Lutz (2002) suggests in a recent review. She outlines an approach that treats emotion as a result of women's existence in the social margins and addresses the issue of emotions and marginality of women as a class. Immigrant women, then, occupy doubly marginalized positions: They are marginal not only by being women but also by their immigrant status and L2 skills. At the same time, drawing on Bakhtin, I argue here that voicing discourses of emotion is not necessarily a sign of weakness or passivity. For example, as Vera described how humiliated she felt by being positioned in a discourse of otherness by native speakers of English, she exclaimed, "No, this is not going to be!" and her tone contained the seeds of resistance and future action. When the women articulate how they feel about a discursive situation, they are not merely expressing an inner mental state. They are engaging in a complex rhetorical relationship with a specific audience and at the same time are expressing a socioideological position. In this vein, as feminist scholar Lorde (1984) suggests, feelings are not simply a reaction to everyday experiences but can serve as a guide to social analysis. Analyzing the emotion discourses of immigrant women can serve as a guide to social analysis in constructing not only theories of gender in SLA, but also feminist pedagogies informed by particular participants in their specific sociohistoric locations.

This study does not suggest that the male participants did not experience feelings or nervousness, fear, or shame. Rather, it shows that the men and the women voiced different discourses in answering their new sociolinguistic realities. Dmitri, for example, did not normally speak of feelings or any discomfort he might have experienced. Indeed, when asked if he ever felt uncomfortable using the L2, he would either deny it or dismiss the topic as immaterial. His wife, however, mentioned that while they worked as servers in the same restaurant, it would be she who would speak to the manager, especially if they had a problem, because Dmitri would get "nervous."

I do not here focus on how the men and women conceptualized gender as a social category; my goal is to show how gender emerges spontaneously in textualized discourse in relation to the L2. In some cases, however, the women overtly commented on their larger roles in the family. Vera, for example, talked of being responsible for helping Aleksei to select appropriate clothes for work or other occasions. She mentioned that she had been doing this throughout their married lives. Vera's appropriating the discourse of responsibility in the family structure is reminiscent of Sylvia's claim that she feels "responsible for understanding" because her misunderstanding could cause some "serious" problem for her family. This sense of responsibility has transferred to the linguistic practices of the couples where by correcting their husbands' linguistic errors, supplying the English words, and providing metalinguistic clarifications for them becomes a way of taking care of the family.

CONCLUSION

Today researchers agree that gender is a socially constructed category that varies across communities and cultures. At the same time, Bakhtin (1981, 1984, 1993, 1994) reminds us that humans themselves are complex sociopsychological constructs that defy formulaic expressions and are difficult to categorize. In pointing out that all individuals retain something that is unique—their own emotional-volitional tone, with their own system of values developed over a lifetime—Bakhtin allows us to move between the domains of social activity and individuated ways of authoring identities. He cautions that seemingly extralinguistic experiences, like one's feelings and desires, are mediated by discourses that are marked by social values and positions. He also reminds us that the formation of subjectivity and consciousness itself cannot be located outside language and is in fact a dialogic phenomenon. The lives of the immigrants in this study and their narrativized worlds provide a context for examining how we answer the voices of concrete and generalized others and author ourselves in a complex interplay of discourses.

Bakhtin's philosophy of language and the self is complex and multifaceted. Although the Russian thinker was not overtly interested in SLA or gender, this article illustrates that his notions of dialogue, discourse, and answerability as inherently social (and yet uniquely individual) phenomena have much to offer to researchers of immigrant communities and the intricate "language-gender interface" (Cameron, 1996, p. 49). His approach allows us to add another dimension, another lens through which to view the gendered self: that of the complex, ever-shifting dialogic relations between the self and the Other that—uniquely to Bakhtin's understanding of languaged subjectivity—are fused with a sense of moral answerability. This sense of dialogic responsibility for the female participants in this study was inseparably bound to the consciousness of others: native speakers of English or their spouses.

The study presented here is only a small step in integrating feminist approaches to gender in SLA with a relational model of viewing the self. Thus, following Cameron's (1996) call for theories that challenge our customary ways of thinking, I suggest that we need more empirical studies in diverse contexts that link critical approaches to L2 gender research with a Bakhtinian prism of subjectivity. As Diaz-Diocaretz (1989) claims, when viewed from a Bakhtinian perspective, language use is no longer seen as a static dichotomy between dominant (male) and nondominant (female) forces, but as "a new, fully dynamic field in which all situations of the word become equally contested and challenged" (p. 135).

ACKNOWLEDGMENTS

In many ways research is a dialogical process. I am deeply indebted to Deborah Hicks for her guidance throughout the project. Our conversations and her insightful comments have greatly contributed to this article. I also thank the two anonymous reviewers and the editor of this issue, Juliet Langman, for their careful attention to the article and constructive feedback.

REFERENCES

Bakhtin, M. (1981). *The dialogic imagination: Four essays by M. Bakhtin* (C. Emerson & M. Holquist, Trans.). Austin: University of Texas Press.

Bakhtin, M. (1984). *Problems of Dostoevsky's poetics* (C. Emerson, Trans.). Minneapolis: University of Minnesota Press.

Bakhtin, M. (1993). *Toward a philosophy of the act* (V. Liapunov, Trans.). Austin: Texas University Press.

Bakhtin, M. M. (1994). K filosofii postupka (Toward a philosophy of the act). In D. A. Tatarnikov (Ed.), *M. M. Bahtin: Raboty 20-h godov* (pp. 9–68). Kiev: Next.

Bartky, S. L. (1996). The pedagogy of shame. In C. Luke (Ed.), *Feminisms and pedagogies of everyday life* (pp. 225–241). Albany: State University of New York Press.

Bergvall, V. L. (1999). Toward a comprehensive theory of language and gender. *Language in Society, 28*, 273–293.

Bergvall, V. L., Bing, J. M., & Freed, A. F. (Eds.). (1996). *Rethinking language and gender research: Theory and practice.* London: Longman.

Butler, J. (1990). *Gender trouble: Feminism and the subversion of identity.* New York: Routledge.

Cameron, D. (1996). The language-gender interface: challenging co-optation. In V. L. Bergvall, J. M. Bing, & A. F. Freed (Eds.), *Rethinking language and gender research: Theory and practice* (pp. 31–53). London: Longman.

Diaz-Diocaretz, M. (1989). Bakhtin, discourse, and feminist theories. *Critical Studies, 1*(2), 121–139.

Eckert, P., & McConnell-Ginet, S. (2003). *Language and gender.* Cambridge, England: Cambridge University Press.

Edwards, D. (1999). Emotion discourse. *Culture and Psychology, 5*, 271–291.

Ehrlich, S. (1997). Gender as social practice: Implications for second language acquisition. *Studies in Second Language Acquisition, 19*, 421–446.

Gergen, M. M., & Davis, S. N. (Eds.). (1997). *Toward a new psychology of gender: A reader.* New York: Routledge.

Hesse-Biber, S., Gilmartin, C., & Lydenberg, R. (Eds.). (1999). *Feminist approaches to theory and methodology: An interdisciplinary reader.* New York: Oxford University Press.

Hicks, D. (1996). Learning as a prosaic act. *Mind, Culture, and Activity, 3*, 102–118.

Holland, D., Lachicotte, W., Skinner, D., & Cain, C. (1998). *Identity and agency in cultural worlds.* Cambridge, MA: Harvard University Press.

Holquist, M. (1990). *Dialogism: Bakhtin and his world.* London: Routledge.

Kaschak, E. (1992). *Engendered lives.* New York: Basic Books.

Leontiev, A. N. (1978). *Activity, consciousness and personality.* Englewood Cliffs, NJ: Prentice Hall.

Lorde, A. (1984). *Sister outsider.* Trumansburg, NY: Crossing Press.

Lutz, C. A. (2002). Feminist emotions. In J. M. Mageo (Ed.), *Power and the self* (pp. 194–215). Cambridge, England: Cambridge University Press.

Morson, G. S., & Emerson, C. (1990). *Mikhail Bakhtin: Creation of a prosaic.* Stanford, CA: Stanford University Press.

Norton, B. (2000). *Identity and language learning: Gender, ethnicity, and educational change.* Harlow, England: Longman.

Pavlenko, A., Blackledge, A., Piller, I., & Teutsch-Dwyer, M. (Eds.). (2001). *Multilingualism, second language learning, and gender.* Berlin: Mouton de Gruyter.

Pavlenko, A., & Lantolf, J. (2000). Second language learning as participation and (re)construction of selves. In J. P. Lantolf (Ed.), *Sociocultural theory and language learning* (pp. 155–177). Oxford, England: Oxford University Press.

Peirce, B. N. (1995). Social identity, investment, and language learning. *TESOL Quarterly, 29,* pp. 9–31.

Piller, I., & Pavlenko. A. (2001). Introduction: Multilingualism, second language learning, and gender. In A. Pavlenko, A. Blackledge, I. Piller, & M. Teutsch-Dwyer (Eds.), *Multilingualism, second language learning, and gender* (pp. 1–13). Berlin: Mouton de Gruyter.

Potter, J. (1996). *Representing reality: Discourse and social psychology: Discourse, rhetoric, and social construction.* London: Sage.

Riggins, S. H. (1997). *The language and politics of exclusion: Others in discourse.* Thousand Oaks, CA: Sage.

Spradley, J. (1979). *The ethnographic interview.* New York: Holt, Rinehart & Winston.

Vitanova, G. (2002). *Gender and agency practices in a second language.* Unpublished doctoral dissertation, University of Cincinnati, Ohio.

Vygotsky, L. S. (1962). *Thought and language.* Cambridge, MA: MIT Press.

Legend of Transcription Symbols

V = Vera
A = Aleksei
S = Sylvia
B = Boris
N = Natalia
D = Dmitri
L = Lydia
P = Peter
G = Gergana
/ indicates a pause
… ellipsis indicates unfinished utterance
[…] indicates deleted text
boldface text indicates an emphasis
italicized text indicates foreign language segments
??? means that the segment is unintelligible
== indicates rapid turn-taking with some overlap

JOURNAL OF LANGUAGE, IDENTITY, AND EDUCATION, 3(4), 279–294

"The Days Now Is Very Hard for My Family": The Negotiation and Construction of Gendered Work Identities Among Newly Arrived Women Refugees

Doris Warriner
University of Utah

Recent studies of second language learning and language contact in bilingual and multilingual communities have been informed by feminist poststructuralist approaches to the study of language, gender, and identity (Cameron, 1997; Cameron, Frazer, Harvey, Rampton, & Richardson, 1992; Eckert & McConnell-Ginet, 1992; Erhlich, 1997; Norton, 2000; Norton Peirce, 1995; Pavlenko & Piller, 2001). Viewing gender as a composite of social and economic relations as well as a set of discursive practices, such research focuses on the various ways that ideologies (of language *and* of gender) mediate those social and economic relations. I utilize such frameworks to analyze the multiple and complicated ways that identities are constructed, negotiated, and deployed within specific situations and circumstances. Examining data from long-term participant observation and recorded interviews, I ask how individual refugee women who have recently arrived in the U.S. are positioned and position themselves as language learners and as immigrants with particular gendered work identities within specific contexts.

Key words: second language learning, identity, gender, refugees, narrative, positioning

In many societal-level discussions of English language learning and the process of refugee resettlement, there is little if any attention paid to the dynamic and unique experiences of women (or girls). Instead, it is assumed that all refugees and all

Requests for reprints should be sent to Doris Warriner, University of Utah, 1705 E. Campus Center Drive, Milton Bennion Hall, Room 307, Salt Lake City, UT 84112. E-mail: Doris.Warriner@ed.utah.edu

English language learners face similar kinds of challenges, constraints, and rewards with the process of second language learning (SLL), becoming bi- or multilingual in an English dominant context, or (in the case of newly arrived refugees) establishing self-sufficiency within the time period allotted by the U.S. government. The complicated experiences of individual women refugees, however, challenge stereotypical descriptions of the experiences of "second language learners," new "immigrants," and "refugees" while illustrating the importance of analyzing the role gender plays in the processes of SLL, refugee resettlement, and the social construction of identity.

In this article, I examine the narrative accounts of three women refugees from Sudan in order to highlight the complicated, hybrid, and multiple aspects of their individual experiences as new immigrants in the United States. The narrative accounts recounted here illustrate the various ways that individual identities are constructed in and through "talk." At the same time, such accounts also serve to comment on the structural (and power) relationships that exist in U.S. society. All of the women in the present study want to find work or to pursue more education or training in order to become more marketable on the job market. Whether they are married or not, the women understand and believe that they should work to improve their own future prospects as well as the situations of their families. Although a "good job" is defined differently by each of the women, none of them is happy with the earnings, hours, or demands of most entry-level (minimum-wage) jobs that they are pointed to. In these ways, the narrative accounts of the women challenge societal-level discourses about the goals, motivations, and abilities of many newly arrived immigrants. English-only advocates, for example, often claim that new immigrants should be provided education and other social services only in English; that new immigrants can learn enough English within three months of arriving in the United States to get a job and become self-sufficient; and that if only new immigrants would study and learn English, they will "make it" in this context. However, waiting lists for English as a second language (ESL) classes, a poor labor market, anti-immigration sentiments, racism, a lack of affordable child care, and difficulties with transportation represent a few of the structural factors that challenge the efforts of new immigrants to establish self-sufficiency and that make such generalizations a myth.

My participant observation and close analysis of interview data with the women highlight the ways that immigrants are positioned by macrolevel discourses about language, language learning, and immigration; the ways that individual women refugees manage to negotiate dominant discourses about the status and importance of English; and the ways that the women's choices with regard to language create particular positions (and thus identities) in relation to dominant-level discourses about language and immigration. In general, the stances of the institution, the teachers, and the students are based on optimistically-held beliefs that newly arrived immigrants will gain access to social and material resources during the resettlement process that will enable them to become involved and socially mobile members of U.S. society. It is believed that they will gain this access through English language learning and the utilization of other resources provided by the adult ESL program in which they are enrolled. Such beliefs constitute local practices that are informed and influenced by larger societal structures and forces such as dominant discourses about the place of new immigrants and the importance of learning English to their effective resettlement. However, as my analysis demonstrates, these women encounter obstacles and barriers in their daily lives that contradict dominant discourses and ideologies about language learning, immigration, and social mobility.

THEORETICAL FRAMEWORK

Theoretical frameworks that rely on poststructuralist and feminist approaches to the study of language, gender, and identity (Cameron, 1997; Cameron, Frazer, Harvey, Rampton, & Richardson, 1992; Eckert & McConnell-Ginet, 1992; Erhlich, 1997; Norton, 2000; Norton Peirce, 1995; Pavlenko & Piller, 2001) foreground the influence of language ideologies and power structures on the options, opportunities, and access provided to second language learners. Such work examines the relationship between SLL and first language maintenance (or shift/loss) and the ways that "paying attention to gender enhances our understanding of agency that may lead to investment in L2 learning" (Pavlenko & Piller, 2001, p. 32). Also, the investigation of local contexts and the community-based ways that men and women are constructed through "gendered practices" (Eckert & McConnell-Ginet, 1992) has permeated the study of language and gender in ways that are useful to the examination of the locally situated construction of individual identities among women second language learners. With a view of gender as a composite of social and economic relations as well as a set of discursive practices, I examine the relationship between SLL and gender with a focus on the various ways that ideologies (of language *and* of gender) mediate those social and economic relations. By examining the role of gender relations in language contact situations and language learning outcomes, it is possible to gain a better understanding of "the meaning assigned by ideologies of language and gender to specific linguistic resources and practices" (Pavlenko & Piller, 2001, p. 39).

In this article, I utilize feminist poststructuralist approaches to investigate the multiple and complicated ways that gendered work identities are constructed, negotiated, and deployed within specific situations and circumstances. Specifically, I ask how the lived experiences of women who are English language learners influence and are influenced by the resettlement process, learning a second language, and the situated, multiple, and dynamic aspects of identity construction. Analyzing data from long-term participant observation and individual interviews with three refugee women,[1] I ask how the women are positioned and position themselves as language learners, as recent immigrants, and as individuals with particular gendered work identities within specific situations and contexts.

I highlight the fact that learning English in and of itself is not enough to enable the women to establish economic self-sufficiency within the U.S. context, in large part due to the gendered aspects of their individual identities. Although the women are able to find entry-level jobs with minimum wage salaries, they find such salaries to be inadequate, and the hours they are able to work at these jobs are constrained by factors such as a lack of affordable and accessible child care and a lack of reliable transportation. Perhaps as a result, the women maintain a desire to further their educations by going to college. The experiences of these women illustrate the consequences of not being able to access certain types of economic and social capital necessary for climbing the American "ladder of success." At the same time, however, the women draw on available linguistic and nonlinguistic resources to cope with the difficulties of their individual situations and, thereby, (re)construct and enact (Wortham, 2001) particular gendered work identities through their actions and their narrative accounts.

METHOD

The data examined here are taken from a larger two-year ethnographic study of the experiences of seven women refugees enrolled in an adult ESL program in the United States. As Eckert and McConnell-Ginet (1992) and others observe, ethnographic methods are uniquely suited to the task of analyzing the ways in which gendered practices influence individual actions, choices, and identities—particularly with respect to making connections between gendered practices and linguistic behavior (p. 485). In the larger study, I conducted more than 100 hours of participant observation and more than 20 individual recorded interviews with women refugees enrolled in an adult ESL program. The data presented here come from the individual recorded interviews.

Drawing on an innovative approach to the analysis of ethnographic data as suggested by Holland and Lave (2001), I analyze the relationship between the construction of individual identity and the influence of historical structures by focusing on locally contentious practices. I accomplish this by examining the specific ways that three refugee women tell stories about work-related goals, issues, or challenges. Following Wortham (2001) and utilizing tools from narrative analysis, I examine storytelling in order to gain a better understanding of the ways the women position themselves in relation to the topic of work, the other characters in the story, the interlocutor (the interviewer), and societal-level discourses regarding language learning, immigration, and resettlement in the U.S. context. I assume that both the content of a story and the way it is told contribute to the representation, construction, and enactment of identity. My analysis illustrates the multiple ways that some of the women construct "working" identities through narrative as well as

the ways that autobiographical narrative accomplishes the interactional positioning (Wortham, 2001) that contributes to self-(re)construction.

I rely on Bakhtin's notions of dialogism (as summarized by Holland & Lave,
2001, pp. 9–14), which claim that (a) people are always in the process of either being
addressed or answering/responding; (b) an individual develops a sense of self
through incorporating "the languages, dialects, genres, and words of others to which
she has been exposed"; (c) dialogic selves are "animated by discourses widely circu-
lating locally and beyond"; and (d) speakers not only use the words of others but also
"take active stances" with respect to those words and, in so doing, enact particular
identities for specific purposes. In these ways, speakers create positions by utilizing
existing language and cultural resources, and their "talk" becomes dialogic by making use of historically influenced structures and processes (such as those expressed
in certain ideologies of language, diversity, and immigration) while also influencing
and structuring the course of history in a locally contingent manner.

I also rely on Sawin's (1999) analysis of the relationship between gender, context, and the narrative construction of identity in which she rejects static,
essentialized conceptions of gender and argues that "any individual's identity is
multiple and constantly re-created as the speaker adopts subject positions in cultural discourses" (p. 241). Sawin reminds us that new trends in feminist
poststructuralist research focus on the ways that women construct and use narrative within particular situations and for specific (sometimes strategic) purposes.
That is, women utilize linguistic resources available to them in order to simultaneously narrate their lives and to construct individual identities within the context
of particular situations and circumstances. As Sawin (1999) notes,

> Each teller develops narrative forms and narrating strategies that work with available
> symbolic and linguistic resources, within and against local social structures and dis-
> courses, and that respond to culturally, historically, and personally determined needs
> and lacks. (p. 254)

Sawin (1999) goes on to emphasize the ways in which narrators use existing resources "to create a personally satisfying and socially acceptable self-presenta-
tion" (p. 254). The emphasis here is on the particulars and complexities of any one
person's narrative accounts as well as the specifics of "strategic situational negotiations," in contrast with earlier attempts to classify women's narratives in terms of
generic or universalizing "models" (p. 254). As the data I analyze here illustrate,
women refugees living in the United States use multiple linguistic resources from
their second language, English, to (re)present themselves in innovative and socially constitutive ways according to the specifics of the situation, while funda-
mentally influencing and shaping personal relationships as well as professional
trajectories in the process. I examine the dialogic character of language in use by
focusing on how the women position themselves in relation to societal-level dis-

courses about language and immigration; in relation to the academic, social, personal and/or professional challenges they face in their daily lives; and in relation to the words or "voices" of others, including the interviewer.

Finally, it is important to consider the influence of the interviewer on the content and form of the narrative accounts. Because I am a native English-speaking researcher and a former employee of the adult ESL program, one might speculate that each of the women was "speaking" or "responding" to me as a representative of the institution. It is possible, and perhaps likely, that the women each foregrounded certain parts of her identity with me (such as her work identity and her assertive, action-taking approach) that she may not have with other interlocutors, particularly those from cultural backgrounds that are unaccustomed to women working outside the home. For example, knowing that I am an American citizen and a White woman, Ayak likely assumed that I would receive her stories about her assertive behavior in a positive and reaffirming way (which I did). In this way, Ayak's construction of identity is social, situated within a specific local context, and dynamic—influenced by and influencing the content and course of the qualitative interview. Let us now turn to Ayak's narrative accounts.

Ayak

Ayak is from Sudan, married, and the mother of two young children. During my first interview with Ayak, she talked about how difficult things were for her family because her husband's job did not provide enough money to cover their expenses. She discussed wanting her mother to come to the United States (from Egypt) to live with her so that she would not have to look for a baby-sitter, and she expressed a desire to find a job to "help" her husband pay the bills. In the following excerpt, Ayak first describes the difficulties of her family's current situation and then draws attention to her plans to work to supplement her husband's income and help her family achieve economic self-sufficiency.

```
 1  A The days now is (.) very (.) hard for my (.) family
 2     for months (.) I send four (.) maybe four hundred dollar
 3     maybe three hundred dollar and a half
 4     I don't have enough money now
 5     my husband (.) he's work alone
 6     no (.) no
 7     me I didn't work
 8  I Right
 9  A Yeah
10     my husband (.) he's have now here run the house here
11     uh, he's run house (.) he's pay the bill (.) he's pay insurance
12     he's, uh, everything ...
```

13   **this is the problem**
14   **I tell my husband 'I (.) I want to looking for another**
15   **babysitter stay with my children (.) I want to work**
16   **looking for a job (.) I help you'**
17   this is a (.) <u>big problem</u>
18   my mother (.) she's need the <u>money</u>
19   and (.) <u>here</u> need the <u>money</u> (.) no, <u>no way</u>
20   **I want to looking for the job**
21 I But now you have a job at the airport?
22 A <u>Yeah</u> (.) I go over <u>there</u> (.) <u>today</u>
23   Yeah (.) they they they give me (.) a paper
24   for the (1.5)
25 I an application?
26 A No, no application (.) they give me paper for <u>direct test</u>
27 I Direct test?
28 A Um hum (.) they tell me you can come (.) at <u>orientation</u>
29   in <u>Tuesday</u>
30 I Good.
31 A **Maybe <u>next week</u> (.) maybe I start <u>work</u> over there**
      (Interview, 4/4/01)
      (See Transcription Conventions on page 294.)

As a whole, this excerpt illustrates the ways that Ayak positions herself in relation to her husband, her mother, the need for child care, and the financial difficulties she faces. It also highlights the ways that Ayak begins to establish for herself an identity in which she imagines the world of work in her future. The topic of the narrative excerpt is Ayak's desire to find a job outside her home to help support her family (here and in Egypt). In the first part of the excerpt (lines 1–12), the rationale for this desire is set up with Ayak's description of the financial challenges experienced by her family, the fact that her husband works "alone," and the fact that she herself "didn't work." In the remainder of the excerpt (lines 13–31), Ayak represents and enacts her newly discovered but very gendered identity as an assertive, problem-solving wife and mother who wants to contribute financially to improving their situation. In this instance, Ayak is both hopeful and cautious, illustrated by her insistence that she wants to look for a job combined with her hesitation and repeated use of the hedge "maybe" (line 31).

In this narrative excerpt, Ayak has utilized available linguistic resources (her limited knowledge of English vocabulary, grammar, and communicative conventions) to begin to establish a gendered work identity. She does this by referring to her husband and to the topic of not having "enough money." The gendered work identity under construction in this excerpt is enhanced in later narrative accounts, illustrated by the following excerpt, taken from an interview I had with Ayak about five weeks later. This passage contains Ayak's detailed description of the conversation she had with the restaurant supervisor during her job interview:

```
 1  A  when I go to interview (.) they ask me 'what you do?'
 2     'why you don't work before?'
 3     I tell him 'because I have small children
 4     I don't have somebody care about him'
 5     They ask me 'what about now?
 6     You get somebody care about him?' I tell 'Yeah
 7     I have my family (.) coming here now (.) in this year
 8     they care about my children' (.) I tell him all that
 9     they tell me 'you need work in the morning' I tell them 'No
10     because I go to school in the morning. I want to learn English
11     more (.) um hum (.) I want to finish (.) uh high school
12     I want to take diploma for high school
13     After that (.) I want to go to college
14     I want to continue (.) my (.) my (laugh) English'
15  I  You told him everything?
16  A  uh huh
17  I  And so he was very impressed
18  A  Yes yes (.) she's very happy
19     She say 'Okay I like you (.) you sound look good'
20     yeah 'you come in next week'
21     uh (.) they give me paper for direct test
22     Yeah I go direct test (.) after that I go (.)
23     they tell me 'tomorrow come and start work'
       (Interview, 5/29/01)
```

In this excerpt, Ayak constructs an identity as a woman who lives with a number of constraints but who is adaptive, resourceful, and able to utilize the cultural knowledge she has gained to improve her situation. In response to questions about why she doesn't have work experience and whether she has anyone to care for her children now, Ayak indicates that, even though she once needed to take care of her small children, she now has a support system in place that enables her to work. Once again, Ayak's narrative account reveals the continuation of a gendered work identity formation.

In Ayak's description of her interaction with the work supervisor, she positions herself as someone who needs a job but also as a mother of young children and as an individual with educational and career aspirations that transcend the limitations of her current situation and its accompanying opportunities. Ayak makes clear that, although she is willing to work hard, she is not willing to compromise the needs of her family or her own personal goals. In these instances, Ayak both represents and enacts her newly constructed but gendered work identity as a woman refugee who is able to find work in order to "help" her family while maintaining her priorities and her personal goals. This is the type of dialogic relationship between individual identity construction and external or historical influences that is central to the framework proposed by Holland and Lave (2001). In the following section, I analyze the ways that a gendered work identity is constructed and displayed through narrative by another woman refugee from Sudan.

Alouette

Alouette is from Sudan and a single mother of eight children (two of whom are nephews that she has adopted as her own). Ever since arriving in the United States, Alouette has struggled to make ends meet. During the time of my interviews with Alouette, this struggle became even more severe and dramatic due to the fact that Alouette had recently put a lot of the money provided to her by the government into an $800 deposit required for renting a house. Alouette rented this house because of the large size of her family and with hopes of establishing some security for her children. During one conversation we had about the struggles associated with paying all her bills, Alouette mentioned that, even though the money provided by the Department of Workforce Services was not enough to make ends meet, she was not optimistic that finding a job would make the situation any better. On the other hand, when I pressed Alouette to think about what kind of job she might like, she said she would like to know where to get training in child care. These topics are illuminated in the following excerpt, taken from our second interview:

```
 1  I    So it's difficult to pay all the bills? (1)
 2  Al   the [power bill] uh (.) is difficult for me …
 3       These here is difficult
 4  I    um hum
 5  Al   but now (.) I don't find (.) a job because when I work
 6       I get five dollars something
 7       I need to learn more
 8  I    Yeah?
 9  Al   like one year (.) one and a half
10       maybe I know the language
11       maybe my money can go
12       something (.) help me like this
13  I    yeah
14       what kind of job do you want to find?
15  Al   when?
16       after?
17  I    after school …
```

18 Al this <u>summer</u>
19   **I go to fill application for child care**
20   **I <u>like</u> the child care perhaps it's=**
21 I   =oh yeah! I think that's a great idea
22 Al yeah
23   **if I <u>found</u> the <u>center</u> for training I go there**
24 I   Yeah?
25 Al **but I don't know where I find <u>this</u> place**
26   **training <u>how</u> to take care of children**
27 I   oh okay
28   you want more <u>training</u>?
29 Al um hum
30 I   on how to take care of children?
31 Al yeah
32 I   but you don't know <u>where</u>
33 Al **yeah I don't know**
        (Interview, 5/4/01)

Here, Alouette says that she cannot support herself and her children on a $5.00 per hour wage (which is apparently what she expects to earn), and so she has decided to stay in school and postpone looking for a job as long as she can. However, she believes it will take at least a year to learn enough English to get a job that will pay more than $5.00 per hour, and this period of time far exceeds government regulations that require refugees to begin supporting themselves after 3–6 months in the United States.

In this excerpt, Alouette constructs an identity as someone who wants to work because she thinks she should but also as an immigrant who questions whether she would be able to earn enough money to handle her expenses. In this way, Alouette ventriloquates arguments about the lack of social support provided to the working poor in the United States, the fact that minimum-wage is not adequate for supporting a family, and the ways that social inequalities are transferred from generation to generation. As such, Alouette's concerns about finding a way to make more than $5.00 per hour (and her attendant pessimism) implicitly critique commonly accepted notions of meritocracy and fairness as well as the arguments and assumptions of English-only advocates. Although Alouette is in the highest level ESL class in this adult education program and has passed a number of standardized tests (and thus has "mastered" English), she cannot find employment that will sustain her family. At the same time, Alouette's narrative illustrates the ways that dominant discourses about the importance of English language learning become internalized by disenfranchised refugees and other immigrants (see e.g., lines 12–17).

Alouette talks about wanting to get training in child care but not knowing where to get this training, saying that "if" there were such a program, she would "go there" (lines 27–34). This part of the excerpt reminds us that one of the challenges articulated by many refugee women with small children is in finding affordable and reliable child care so they could study English, go to community college, or find a "good" job. Alouette, like other women in this study, is clearly aware of the fact that a large population of immigrants and refugees needs such services, and she envisions herself as part of the solution to this widespread problem. In this way, Alouette positions herself within her narrative account as a person with a future job in the child care industry and as a resource to other women in similar situations. In so doing, Alouette imagines a gendered work identity within the constraints of her educational and life experiences.

María

María is from Sudan and married with two children. She has years of training and work experience in accounting in her home country, and her husband works for a window manufacturing company. María would like to go to college for more training in accounting and "more education" in general, but she would also like to find work to contribute to family expenses and to establish a life here in the United States. During our first interview, María discussed why she would like to find a job.

```
 1  M  Before I came to school
 2     staying home is really difficult because I want to work
 3  I  Yes.
 4  M  Yeah (1)
 5     and when my husband start to work
 6     uh we don't have any assistance because we
 7     we we are not given (.) anything
 8     from Workforce
 9  I  Why?
10  M  they say that my husband's working
11  I  oh, okay
12  M  yeah
13  I  so he's working so you can receive no assistance
14  M  yes
15  I  does he receive enough money for your ?
16  M  it's it's not enough
17     I don't know why
18  I  Oh. It's not enough?
19  M  yeah
20  I  but they give you no assistance?
```

21 M   no assistance
22   **even up to now we don't have**
23   **insurance**
24   we don't have Medicaid
25   you know
26 I   no Medicaid
27 M   no Medicaid for children
        (Interview, 4/3/01)

In this excerpt, María provides her reasons for wanting to work and describes the financial difficulties facing her family. She begins by saying that "before I came to school, staying home is really difficult because I want to work" and then explains the details of her family's situation in more detail. Like Ayak, María would like to work in order to supplement her husband's income. A few minutes later, we discuss the plans María has for finding a job, and she tells me she will begin to look seriously after she completes the next session (when she will have completed the required credits and tests to graduate from the program).

In this excerpt, María positions herself as a woman/mother who would prefer to work but who will go to school if she cannot work (and to avoid staying at home). In her description of the situation, she highlights the fact that, because her husband is employed, they receive no additional assistance and cannot afford health insurance among other things. She explains (with exasperation) that she does not know why her husband's income is not enough to live on, but it is not. In this presentation of the difficulties of her personal situation, María articulates a larger social and political concern with the forms of assistance provided to refugees (and immigrants in general) during the resettlement process and ventriloquates complaints often made about current government policies. Although her husband has found work in the United States, his earnings are not enough to cover the expenses faced by his family or to establish self-sufficiency. To the contrary, the structures in place serve to penalize his efforts rather than reward them.

The following excerpt highlights María's willingness and desire to find a job while drawing attention to the structural forces and obstacles that continue to challenge the efforts of her family to establish themselves here in the United States. Even though María is proficient in English, has completed the equivalent of a high school diploma in the United States, and has training in accounting, she does not have anyone to care for her children while she applies and interviews for jobs.

1 I   And (.) when will you look for a job?
2 M   After (.) **after we have (.) completed this (.) course**
3   **I want to look for job**
4   I started giving my application (.) now
5 I   Really?

```
 6  M   Yeah
 7  I   So you will start next week looking for a job?
 8  M   Yeah (.) I started already
 9  I   you started already
10  M   yeah because I fill
11      my husband bring me application to fill and give it (1)
12  I   what kinds of applications did you fill out?
13      what kinds of jobs?
14  M   I like uh (.) cashiering
15  I   cashier?
16  M   yeah or selling thing
17  I   or selling things?
18  M   yeah
19  I   so you want to work in a store?
20  M   yes
21  I   uh huh
22      any other applications?
23      did you fill out anything else?
24  M   I filled that one in (.)
25      post office (2)
26  I   good
27      is it difficult to fill out the job applications?
28  M   it's not too difficult
29  I   not too bad?
30  M   yeah
31  I   have you ever had job interview?
32  M   here?
33  I   here?
34  M   no
35  I   no not yet?
36  M   yeah not yet
37      because I have need to stay home with these children
38      yeah
        (Interview, 4/3/01))
```

In this excerpt, Moría talks about the work she believes she is able to do as well as the challenges she faces in trying to find employment. She says she would like to be a cashier or a sales clerk in a store and that she has filled out an application to work at the post office. When I ask Moría if it is difficult to fill out the applications, she says "it's not too difficult." However, when I ask her if she has ever had a job interview here in the United States, she says she has not "because I have need to stay home with these children." Moría has positioned herself as someone who would like to work and to further her education and/or training but who has not yet had a job interview because she needs to care for her two young children. In this way, the gendered work identity María constructs through narrative resembles the identity established by Ayak as both women want to work to contribute to household expenses but are challenged to find reliable child care for their young children. Like Ayak, María is resourceful; she tells me that she and her husband are going to "exchange"—they plan to find jobs that require them to work at different times of the day so that one of them is always available to care for their children.

CONCLUSION

In this article, I described the different ways that three refugee women from the Sudan use linguistic resources and their growing cultural knowledge to construct individual gendered work identities within particular local contexts. My analysis of vignettes from the women's narrative accounts illustrates the ways that positioning is achieved through talk, the complicated and locally situated ways that people construct distinct identities through storytelling and improvisation, and the various ways that particular women refugees utilize linguistic resources in their second language (English) to construct or transform their identities within the context of the local situation. The women's narrative accounts highlight the locally contested aspects of their individual identities while illustrating the dynamic and gendered processes associated with refugee resettlement and English language learning in the U.S. context.

I have also explicated the complicated ways that Ayak, Alouette, and María construct multiple and complex identities for themselves through autobiographical narrative by utilizing particular linguistic devices in their second language (English) and their growing knowledge of the culture and traditions in the United States. The resulting hybrid identities displayed by the women in their autobiographical narrative accounts combine the traditional roles of wife and mother with the responsibilities of securing employment and becoming advocates for their families in this new and unfamiliar context. In this way, their narrative accounts both represent and enact gendered work identities that are specific to the circumstances of their lives in the United States.

In addition, I have examined the ways that historical structures such as ideologies about language, language learning, and immigration influence and are influenced by the complicated and contradictory nature of individual practices and beliefs. For example, while the women in this study accept some dominant ideologies regarding English and SLL (such as the ideology that connects the acquisition of English with getting a job and improving one's situation), their experiences provide a harsh critique of this ideology. Even though the women in this study are considered very successful students within the context of this particular adult ESL program (completing

various levels of English language learning as well as the requirements for comple-
tion of their GEDs), they are limited in their abilities to use their English language
learning skills and knowledge to deal with everyday situations and obstacles let
alone with the occasional major setbacks they encounter. In addition, in spite of the
fact that they have obtained (or will soon obtain) credentials or have completed ex-
aminations that certify their attainment of a high school diploma and mastery of certain subjects in English (reading, writing, and math), they lack the financial resources or social networks to put those institutional credentials to use in finding what
they consider to be a "good job." To achieve true self-sufficiency, the women's ability
to understand and speak English must be supplemented with the communicative
competence and cultural capital needed to know where to get information, how to
ask for it, and what to do with that information. Advocates of English-only policies
and practices, however, often ignore the fact that additional resources are required to
create and sustain productive lives in this new context.

On the other hand, it is neither accurate nor fair to portray the women in this study as
helpless victims of unfortunate circumstances, anti-immigration attitudes, or English-only policies. On the contrary, through their creative and strategic use of available
resources and with an increasingly sophisticated understanding of "how things work"
here in the United States, the women act in ways that challenge stereotypical notions of
the poor, failing, unmotivated immigrant. Finding ways to balance motherhood,
school, and work, a number of women in this study demonstrate their firm resolve to
make it here in spite of the numerous obstacles they face. By positioning themselves in
particular ways, the women construct and reconstruct particular identities for themselves as working mothers, committed students, and people with goals and dreams. In
this way, their testimonies illuminate the ways that their experiences as women refugees serve as stories of triumph over the obstacles and hardships of their disadvantaged
circumstances. My analysis of the relationship between their experiences and historical influences sheds light on the complicated ways that enduring struggles such as
those faced by these women and individual identity construction are mediated, influenced and transformed by local and sometimes contentious practices.

ACKNOWLEDGMENTS

I am indebted to the women refugees who talked with me and shared their concerns, anxieties, and hopes. I am also grateful to the teachers and administrators at
Valley Instruction and Training Center (a pseudonym) for supporting my research.
I acknowledge Nancy Hornberger, Stanton Wortham, and Sofia Villenas, who all
provided invaluable guidance and feedback while I worked on conceptualizing and
writing up the larger dissertation study; and I thank Bryan Brayboy and Anne
Pomerantz, who each read earlier versions of this article and made very useful suggestions. Finally, I appreciate the recommendations of two anonymous reviewers

and Juliet Langman, the editor of this special issue, during the preparation of this article.

ENDNOTE

[1]Although the women refugees in the larger dissertation study come from different ethnic and national backgrounds (e.g., Sudan, Bosnia, and Iran), I focus on the experiences of three women from Sudan in this article.

REFERENCES

Cameron, D. (1997). Theoretical debates in feminist linguistics: Questions of sex and gender. In R. Wodak (Ed.), *Gender and discourse* (pp. 21–36). London: Sage.

Cameron, D., Frazer, E., Harvey, P., Rampton, M. B. H., & Richardson, K. (1992). *Researching language: Issues of power and method.* New York: Routledge.

Eckert, P., & McConnell-Ginet, S. (1992). Think practically and look locally: Language and gender as community-based practice. *Annual Review of Anthropology, 21,* 461–490.

Erlich, S. (1997). Gender as social practice: Implications for second language acquisition. *Studies in Second Language Acquisition, 19,* 421–446.

Holland, D., & Lave, J. (Eds.). (2001). *History in person: Enduring struggles, contentious practices, intimate identities.* Santa Fe, NM: School of American Research Advanced Seminar Studies.

Norton, B. (2000). *Identity and language learning: Gender, ethnicity and educational change.* London: Longman.

Norton Peirce, B. (1995). Social identity, investment, and language learning. *TESOL Quarterly, 29,* 9–31.

Pavlenko, A., & Piller, I. (2001). New directions in the study of multilingualism, second language learning, and gender. In A. Pavlenko, A. Blackledge, I. Piller, & M. Teutsch-Dwyer (Eds.), *Multilingualism, second language learning, and gender* (pp. 17–52). New York: Mouton.

Sawin, P. E. (1999). Gender, context, and the narrative construction of identity: Rethinking models of "women's narrative." In M. Bucholtz, A. C. Liang, & L. A. Sutton (Eds.), *Reinventing identities: The gendered self in discourse* (pp. 241–258). New York: Oxford University Press.

Wortham, S. (2001). *Narratives in action.* New York: Teachers College Press.

Transcription Conventions

____	stressed word or syllable
(xxx)	speech inaudible or hard to discern
bold	speech of central interest to analysis
(.)	pause of less than one second
(1)	silence/pause timed to the nearest second
?	rising intonation
.	falling intonation
,	pause or breath without marked intonation
=	two utterances closely connected without noticeable overlap

JOURNAL OF LANGUAGE, IDENTITY, AND EDUCATION, 3(4), 295–311

"I Always Had the Desire to Progress a Little": Gendered Narratives of Immigrant Language Learners

Julia Menard-Warwick

University of California, Davis

Based on a qualitative study of Latin American adults enrolled in a California English as a second language (ESL) program, this article examines the ways in which gender as a social construct mediates learners' decisions and opportunities to learn English. In narratives audiotaped during life-history interviews, participants shared their varying responses to the gender ideologies and practices of their communities, and their understanding of how these responses affected their language learning. This article analyzes the narratives of two adult immigrant ESL students to better understand learners' perspectives on the connection between gender identities and second language learning. It concludes by discussing the implications of this research for adult immigrant ESL programs.

Key words: narrative, gender, second language learning, immigrants

This article explores narratives on gender and second language (L2) learning told by two Latina immigrants to the United States, both of whom were participants in a larger qualitative research study conducted at an English as a second language (ESL) family literacy program in an urban California neighborhood. Although study participants saw themselves positioned in multiple ways by their communities and by the larger society, gender identities emerged as a key factor in many learners' decisions to pursue or not to pursue English language competence at particular times in their lives. In interviews, different adults interpreted the cultural imperatives of gender in varied ways, depending on their histories and present circumstances.

Requests for reprints should be sent to Julia Menard-Warwick, Department of Linguistics, One Shields Avenue, University of California, Davis, CA 95616. E-mail: jemwarwick@ucdavis.edu

This article explores the participants' viewpoints on the gendered practices and ideologies of their own communities, along with the ways they portray themselves in narrative as living out or resisting these practices and ideologies. In so doing, this article examines participants' perspectives on the ways their gendered identities as immigrants in the United States have mediated their learning of the English language. In discussing these issues, I define *ideologies* as systems of meaning generated in power relations (Fairclough, 1992) and *practices* as recurrent, goal-directed, socially meaningful human activities (Scribner, 1997).

SECOND LANGUAGE LEARNING AS GENDERED PRACTICE

If identity is seen as "the social formation of the person ... through complex relations of mutual constitution between persons and groups" (Wenger, 1998, p. 13), the negotiation of gender is clearly an important part of this process. As Eckert and McConnell-Ginet (1992) write, "Gender is constructed in an array of social practices within communities, practices that in many cases connect to personal attributes and to power relations but that do so in varied, subtle, and changing ways" (p. 484). Susan Ehrlich (1997) draws on these authors' formulation to make sense of the influence of gender identities on L2 learning, arguing that men and women in particular communities often have different attitudes towards second languages and unequal opportunities to learn and use them, "depending on the particular way that gender identities and gender relations are constituted in that community" (p. 430).

Many qualitative studies of language learning in immigrant communities support Ehrlich's contention that gendered practices within social groups have a strong influence on individuals' opportunities to engage with the target language (e.g., Cumming & Gill, 1992; Goldstein, 1997). In examining the connection between literacy, L2 learning, and gender among Latina immigrants in California, Rockhill (1993) found that while men in this community had more chances to speak English in the workplace, women did most of the English-language literacy work of the household. At the same time, many women were frustrated in their desire for further education in English due to family pressures, including domestic violence.

Gendered practices and identities are often contested in immigrant communities (Kibria, 1993; Schecter & Bayley, 2002), with individuals and families making choices about which aspects of traditional gender roles they will struggle to maintain and which they will (sometimes happily) let go. Moreover, while ethnographic studies demonstrate that the practices of particular communities are an important part of the context for adult literacy and language development, Goldstein (1997) points out that to understand how "human agency interacts with people's social roles, relationships, and goals," it is necessary to examine the stories of particular individuals (p. 177).

To account for this interaction between agency and social context, Norton Peirce (1995) contends that individual learners make an "investment" in the target language "with the understanding that [in so doing] they will acquire a wider range of symbolic and material resources" (p. 17). In autobiographies, immigrant women themselves make clear that L2 language and literacy development is intimately tied to their gender-mediated struggles to meet their personal goals in a new land (e.g. Alvarez, 1998; Hoffman, 1989). In a discussion of gender in language learning memoirs, Pavlenko (2001) cautions that all such narratives are "discursive constructions" rather than "factual statements" (p. 214). She notes that most autobiographers are middle-class and highly educated, and wonders whose stories remain untold. Nevertheless, I concur with Pavlenko that immigrants' stories contribute to a "complex, theoretically, and sociohistorically informed examination of individuals' perceptions of the social and cultural contexts of second language learning" (p. 214), including the influence of gender.

While Pavlenko studied published memoirs, it is possible through oral interviews (and ESL classroom activities) to explore the perceptions of individuals who are unlikely to ever write an autobiography. Oral and informal written narratives can elucidate diverse learners' insights on their positioning within families, communities, and society and on how this positioning has facilitated or impeded L2 learning over time.

NARRATIVE: MAKING SENSE OF EXPERIENCE

As texts that draw connections between events over time, narratives reflect human experience in the light of typical human concerns. Given the importance assigned to gender within most human cultures, gendered themes are typical within a variety of narrative genres. Indeed, narrative offers a linguistic means of constructing particular kinds of identities, including gender identities (Bucholtz, Liang, & Sutton, 1999), and is thus "a crucial resource for socializing emotions, attitudes and identities, developing interpersonal relationships, and constituting membership in a community" (Ochs & Capps, 1996, p. 19).

At the same time, narrative is a way to reflect on and make sense of experiences and identities. Bruner (1990) argues that people tell stories about "trouble" as a way to handle and explain problematic circumstances, to themselves and to interlocutors. Narratives of the past often reference current concerns, implying a connection between past and present (Ochs & Capps, 1996). In this way, narratives make sense not only of "trouble," but of significant life change. Narratives of immigration and language learning cited above (e.g., Hoffman, 1989) certainly fulfill this function, allowing immigrant autobiographers to explore both the pain and the value inherent in building an identity in a new language and culture.

Indeed, the *evaluation* of experience is a crucial function of narrative (Labov, 1972). Beyond creating psychological and cultural coherence, narrative is a means to share or impose a point of view or moral stance (Bruner, 1990; Ochs & Capps, 1996). For example, immigrant autobiographers draw upon their experiences to take stands against linguistic and ethnic discrimination on the part of the Anglo majority and against sexism in their own communities (e.g., Alvarez, 1998). Narratives recount human experience in a way that is not necessarily factual but that is true to the teller's perspectives on that experience and shaped by the teller's perception of the listener(s) (Cortazzi, 1993). In this article I discuss interview narratives recounted to me by two Latina immigrants in the United States that detail their perspectives on how their L2 learning has been mediated by the varied ways they have lived out or resisted the gendered practices and ideologies of their communities.

METHOD

I spent 7 months during 2002 as a participant-observer in a California ESL family literacy program, which I will refer to as the Community English Center (CEC), volunteering as a teacher's aide. Most of the students in these classes were immigrant women from Latin America. As much as possible I became an active member of the ESL program community at the CEC through volunteering. I tutored students individually, led small groups, answered questions on a variety of grammatical and cultural topics, and occasionally translated when students were confused.

In addition to my classroom participation, I conducted audiotaped life history interviews in Spanish with eight students, one man and seven women. These participants came from Mexico, Guatemala, El Salvador, Nicaragua, and Peru. To some extent I selected particular interviewees because we had already developed some rapport during classroom activities; at the same time, I made an effort to choose participants that would be representative of CEC students in terms of gender, ages of children, level of education, work status, country of origin, and number of years in the United States. I told participants that I was interested in understanding the connections between their lives outside the classroom and their experiences of learning English in the classroom.

Conducted in the homes of participants, interviews were almost entirely in Spanish, although at times participants codeswitched into English in order to demonstrate their use of the language in community settings. In introducing these interviews, I encouraged participants to tell stories about their experiences while answering my questions. During the interviews, I asked them about childhood experiences with literacy outside of school; experiences with formal schooling; the educational, literacy, and work experiences of family members; decisions about schooling, work, immigration, marriage, and children; work histories in their homelands and the United States; personal experiences with reading and writing; experiences learning English in and out of school; and goals for the future. These interviews were then transcribed by a native speaker of Spanish. I began open, thematic coding (Bogdan & Biklen, 1998; Boyatzis, 1998) after collecting and reading through all of my data. Larger themes of *social positioning* and *learning* emerged through this process. Once I finished the thematic coding, I pulled out all the narrative data segments corresponding to subthemes of social positioning and learning, such as gender, immigration, and literacy, which I had identified as important in my reading of the data.

This article examines gendered themes that appear in life history interviews with two of these focal participants, Camila from urban El Salvador and Trini from rural Mexico.[1] I chose to focus on these participants in this article because they took the most strongly contrasting positions on gender issues. I conducted four interviews with each of these women, ranging in length from 1½ to 2½ hours. Here I present selected narratives concerning the participants' perspectives on the gendered practices and ideologies of their own communities and the ways their gendered constructions of identity as immigrants in the United States had mediated their learning of the English language.

While these are stories of participants' "real-life" experiences, readers should note that the stories have twice undergone a shaping process to meet the demands of an audience (Cortazzi, 1993). First, the research participants constructed their accounts according to their interpretations of my identity as an Anglo teacher. Second, I have selected from many hours of audiotaped interviews those aspects of participants' stories I see as most relevant to the connections between gender and language learning in immigrant communities. In so doing, although the responsibility for selecting and interpreting the narratives is mine, I have tried to represent the tellers in keeping with the ways that they represented themselves to me.

In sharing my findings, I begin with background information on each of the participants. I then share narratives about their construction of gender as young people. Through further narratives, I illustrate their perspectives on how these early constructions of gender led to later experiences of immigration, work, and L2 learning.[2] All information about the participants is as of the time of my interviews with them in 2002–2003.

CAMILA'S NARRATIVES

Camila, a 34-year-old Salvadoran with a high school education, had lived in California since 1989. Camila and her husband Marcos, an electrician, had a sixth-grade son and a second-grade daughter. In the fall of 2002, she was not working outside the home and had no immediate plans to do so. Camila was in the intermediate class at the CEC, where she had studied for 2½ years.

"I Understood That My Position Was Less Than Theirs"

In the mid-1980s, Camila graduated from high school in war-torn El Salvador and immediately entered a teacher's college. However, she said that she only studied for 1 year, because at that time, "The soldiers were going around killing teachers … so that was the fear that I felt" (Interview, 11/14/02). After dropping out of school, Camila moved in with the family of her boyfriend Marcos (now her husband). However, when Marcos left to seek work in the United States, she felt uncomfortable in his family's home and went back to live with her mother. It was 2 years before Marcos could send money for Camila to follow him. Meanwhile, her mother, sister, and teenage nieces treated her like an abandoned mistress:

> My mom let me know that my position in the house was no longer the same as it was when I was a *señorita*. … I felt humiliated, because my nieces and my little sister were still *señoritas*, and I wasn't anymore. … Because I understood that my position was less than theirs. (Interview, 12/5/02)

To get out of the house, Camila took a job as a live-in maid for a neighbor: "I felt useful, not like a kept woman. And I didn't feel the scornful glances [*miradas discriminantes*] of my nieces or my sister." However, this job created new problems for Camila. The neighbor was earning extra money by cooking meals for unmarried men. The regular arrival of these men at the house led to vicious rumors, which even reached Marcos in California. In an angry letter, he accused her of "prostituting" herself with the men who visited the house and demanded that she move back in with her mother. Camila did so and found a job in a furniture store. However, one day when she was alone sweeping out a storage room, her boss approached her and began asking her questions about her boyfriend in the United States. When he suggested that perhaps her boyfriend had found someone else, Camila looked offended, and he came closer as if to apologize, but then he grabbed and embraced her:

> And when I grabbed his hand to take it off me, he spun me around, and he grabbed me and kissed me. So when I felt him kiss me, I remember that I pushed him away with fury … "what's happening with you, you dirty old man?" (Interview, 12/5/02)

Camila quit that job and did not look for another. She lived with her mother until Marcos sent money for her to join him in California. While she had had three jobs in the United States, baby-sitting, sweeping up in a hair salon, and stocking shelves in a department store, none of them had lasted long due to conflicts with employers or coworkers. Aside from one period of unemployment in the early 1990s, Marcos had worked steadily as an electrician, and Camila had been a homemaker. Until recent years, most of her L2 learning took place within the family home.

"You Are Never Going to Learn English"

Describing her life in the early 1990s when her son was small, Camila said that she rejected the option of taking English classes at a college near their home because tuition was $70 a semester, plus the price of books:

> So I thought that is less money for my husband's dream [of buying a home] or for my son's education. So that's why I never demanded [*exigí*] to go to the college, and I always got by with the television. (Interview, 1/9/03)

Once she put her mind to it, Camila in fact learned considerable English from the television. However, during her first few years in the United States, she primarily watched Spanish-language soap operas (*novelas*) and had no desire to change the channel. English "sounded ugly" to her, and she did not feel much necessity to learn it. In her urban neighborhood, daily needs could be satisfied in Spanish. It was not until the mid-1990s that Marcos finally persuaded her to take English seriously, in a single dramatic confrontation. One night when she was watching *novelas,* he came up to her and grabbed the remote control. When she demanded to know what he wanted, he asked her in a quiet voice if she loved their son, Marquitos, then 3 years old:

> "What I want to tell you is this," he said. "One day ... the boy is going to grow up," he said. "And you are never going to learn English. And what's going to happen ... is that when he is a **teenager** [English in the original] ... close your eyes and put this photograph in your mind," he said, "you are going to be sitting right here where you are, watching your soap operas in Spanish, and next to you he is going to be sitting with the telephone, speaking English and making drug deals. ... When you ask him what he is doing, he is going to think, 'My mom is a fool, she doesn't speak English.'" ... And I looked at my son, and I felt a lump of tears in my throat [*un nudo en la garganta de llorar*]. (Interview, 1/9/03)

Camila said that was the last night she watched television in Spanish. She began learning English by paying attention to the numbers on Wheel of Fortune, translating them in her mind into Spanish. Around this time, she began taking her children to the neighborhood library and checking out books in English to read aloud to them. However, she did not start attending classes in English until the year 2000, when the family succeeded in buying a home, and she found out that free ESL classes were offered at the CEC, two blocks from her children's new school, during the hours when they were in class.

TRINI'S NARRATIVES

Trini, a 34-year-old Mexican from rural Michoacán, had lived in California for 15 years. She had a sixth-grade education. Trini's husband Alfredo also came from Michoacán, but she met and married him in California. During the interviews, Alfredo worked in a mattress factory, while Trini cooked hamburgers in a fast-food restaurant. They had two children, a son in kindergarten and a daughter in pre-school. Trini had learned conversational English at work but began studying ESL in the beginning class at the CEC in the fall of 2002.

"I Always Had the Desire to Progress a Little"

When Trini began studying at the primary school in her village, it only went up to third-grade. After her third-grade year, some new classrooms were opened, and she was in the first class of children who were able to complete sixth-grade. Trini's father worked in construction, and her mother sold handicrafts, but the family was poor. They could not afford to send their children on the bus to secondary school in the next town. Moreover, Trini's father feared for his daughters' virtue, saying,

> The [high school] girls are real clowns, they just go, and you don't know what they are going to do, and in the end they don't even study, in the end they are going to come back pregnant. (Interview, 12/14/02)

Despite the family's poverty, he was likewise reluctant to let his daughters out of the home to work for pay. Trini's sisters were obedient in this regard, but Trini was not:

> My dad didn't let us [*no nos dejaba*] work in the fields, and I went without his permission [*sin su permiso*]. I went to pick strawberries, to pick toma-toes, but way out in the country, and my sister never went. (Interview, 12/13/02)

Trini explained her disobedience to me by reiterating the poverty of her family:

> We didn't even have bread, at times we didn't even have a school notebook like I told you, at times we didn't even have a pencil. So I always had the de-sire to progress [*progresar*] a little. (Interview, 12/13/02)

In keeping with this desire for "progress," Trini immigrated to the United States at the age of 18, following her older brothers. Soon after arrival, she met and married Alfredo, who initially shared her father's old-fashioned values. When Trini found a factory job, conflict erupted:

> I came home, and I was really happy, and my husband comes in, and I talk to him. "No, you're not going to work," and I did the same thing that I did with my dad. I went without. … I said, "Ah, I'm going to go anyway." And I did go. He went to work, like at five, and I left, like at five or six, just a little after him, because I had to be there at seven. … And I spent like an hour walking. (Interview, 1/17/03)

After about two weeks, Alfredo relented and began to give Trini rides to work. She told me that this was a difficult time, that he was angry at her for her independent attitude, but she persisted. Her working allowed them to thrive economically even during periods when Alfredo was unemployed; after about seven years of marriage, she said, he changed his ideas about men and women and was now supportive of her endeavors.

"So That They Also See That I Am Making a Little More Effort"

Trini had begun learning English soon after arriving in the United States, at her first fast-food job. The manager, who did not know Spanish, would show her how to put the burgers together, explaining each step in English. She learned vocabulary like pickles and cheese by following directions.

However, Trini also had a chance to pick up speaking skills in English at work through the practice of workplace teasing. At the restaurant, most of her coworkers were Asian immigrants, and she had developed friendly relationships with them through the medium of English. She seemed to have particular fun teasing and being teased by a Laotian woman, Angie, the girlfriend of the restaurant's manager, Kevin. When I asked her about practicing English at work, Trini was able to quote both herself and her coworker in English. In fact, the entire passage below was in English in the original except for the first three words:

> I ask her, "hey, Angie, you and Kevin uh uh have a seven years boyfriends? And no married! [laughing] Why you … a long time no good!" And she say, "yes is long time, my life is no … nothing important, nothing especial, nothing different, everyday same thing. When I stay work, he stay in the house, when he here in the Burgerland, I stay in my home, never going to some place, never we go to dancing, [laughs] never to the theater." And I tell her,

"Oh, look for one more, no more Kevin," [laughs] and she say, "Oh! I tell Kevin what you say!" (Interview, 11/8/02)

For many years, the English that Trini picked up in the workplace was enough for her to meet the demands of daily life. After her children were born, however, Trini felt differently. With her daughter in preschool and her son in kindergarten, she saw both the possibility and the necessity of pursuing English classes herself:

My son was asking me a lot of words in English and I don't know what to tell him. ... His homework is easy but for me it's difficult because I don't know, right? ... And so now I began to think of them ... that maybe if I begin [to learn English], like to show them a little ... that I want to get ahead [*salir adelante*], so that they also see that I am making a little more effort. (Interview, 1/17/03)

Fond of reading both magazines and storybooks in Spanish, Trini did not feel that she could read English at all before starting classes at the CEC. After 2 months of daily study, however, she reported to me that she could now read the words that she knew how to say. Explaining how her husband's ideas about women had changed over the years, she noted that he was supportive of her studies: "And he, a little, is encouraging me, and ... at the beginning, when we got married, he wouldn't have let me" (Interview, 1/17/03).

DISCUSSION

As the narratives of Camila and Trini illustrate, L2 learning is not so much mediated by "the way that gender identities and gender relations are constituted in [a] community" (Ehrlich, 1997, p. 430) but rather by the way that individuals *respond* to the gendered expectations that are placed on them by their families and communities. Although these women were similar in many ways (native language, age, marital status, length of time in the United States), lived in the same neighborhood, and studied in the same ESL program, they had grown up in very different settings: rural Mexico in peacetime and urban El Salvador in wartime. They came from different communities with similar gender ideologies but somewhat different gender practices.

Based on their narratives, both women came from communities where an ideology of male authority over female family members was widely accepted. Trini portrayed both her husband and her father making decisions about her education and work without consulting her. A repeating phrase in her narratives is "my dad didn't let us go [*mi papá no nos dejaba ir*]," a phrase echoed in her husband's initial response to her finding a job: "no, you're not going" [*no, no vas*]." The verb "*ir* (to go)" is at the center of Trini's reported arguments with both of these men, indexing their attempts to restrict her mobility. Camila similarly depicted her husband deciding what was best for her and the family: he was the one who chose to immigrate to the United States, and he was the one who eventually decided that it was time for her to learn English.

Both women also described themselves as having primary responsibility for the care of their children: Trini was the one who had to answer her son's questions about English, while Camila, not her husband, needed to monitor their teenager's phone calls. In addition, both narrators reported familial concerns about their sexual virtue. Camila depicted her female relatives' negative reactions to her living with her boyfriend. Trini described her father's fear that his daughters would come back pregnant if he sent them to secondary school. In these stories, women's sexual activities outside marriage were seen as a source of shame for their families.

However, in the realm of practice, the community Trini described was more restrictive of young women's mobility than Camila's. In her interviews, Camila spoke at great length about her decision to attend an academic high school and then to continue on for teacher training. At no point in these narratives did she voice concern on the part of anyone that higher education might threaten her virtue in the way that Trini's father feared. Moreover, while specific jobs in Camila's narratives affected her reputation or exposed her to harassment, no one in her stories questioned the idea of a woman earning money outside the home. In contrast, Trini had to disobey both her father and her husband to work outside the home at all. Moreover, her father prevented her from attending secondary school, and she said that her husband early in their marriage would not have allowed her to attend English classes.

However, in contrast to Trini, Camila tended to take for granted traditional gender ideologies and practices. In interviews, she delineated the gendered constraints that she had felt in her life, without questioning them or rebelling against them. When her fiancé emigrated to the United States, leaving her no longer a *señorita*, she describes herself as unable to protest the "scornful glances [*miradas discriminantes*]" of her sister and nieces, because she says repeatedly that she felt herself to be "less than them [*menos que ellas*]," "humiliated [*humillada*]," and even "like a piece of trash [*como una basura*]." This language of AFFECT and JUDGEMENT (Martin, 2000) suggests that Camila accepted her family's moral authority and their negative evaluations of her behavior.

Camila's sense of vulnerability appeared to have been exacerbated by her wartime experiences. In most of her narratives, the world around her was a place of danger, and she was in need of male protection. She used two words for fear, "*temor*" and "*miedo*," a total of 23 times in four interviews (in four interviews of similar length and on similar topics with Trini the weaker word "*miedo*" appears four times, while "*temor*" does not appear at all). In this way, Camila exemplified Bruner's (1990) claim that telling narratives is a way to deal with "trouble." In these stories, she went along with gendered constraints because struggling against them left her vulnerable to danger. As she voices a friend saying in one wartime narrative, "well, there's nothing else you can do."

Trini had been more inclined to resist her community's gender practices and ideologies. Many of her stories were "counternarratives," accounts that resisted the status quo in her family (Ochs & Capps, 1996). In these stories, Trini took a moral stance (Bruner, 1990) against traditional gender expectations. Whereas many authors (e.g., Cumming & Gill, 1991; Rockhill, 1993) have tended to see gender roles as entrenched and inflexible, Trini's struggles exemplify the potential fluidity of family life in immigrant communities (Kibria, 1993; Schecter & Bayley, 2002). Depicting her male relatives' attempts to restrict her mobility, Trini repeatedly highlights her own agency in response: "I went without permission [*sin permiso*]." In Trini's self-portrait, she saw no reason to put up with poverty when she could earn money for herself: "I always had the desire to progress [*progresar*]." Whereas Camila said over and over that she felt "humiliated [*humillada*]," Trini told me repeatedly of an opposite affective reaction (Martin, 2000): She did not feel that she was doing "anything bad [*nada malo*]." It is the case, however, that Trini's most significant battles had been conducted around the issue of working for pay; her stories did not challenge traditional sexual morality or question her primary responsibility for child care.

Despite the differences between these two women, both depicted themselves as committed wives and mothers, *invested* in their familial roles (Norton Peirce, 1995). Their language learning endeavors had arisen from these primary investments. Portraying herself in narrative as a wife and mother who needed to place her own needs behind those of her family, Camila rejected "education and advancement" for herself as a threat to family unity (Rockhill, 1993). She said that she never "demanded [*exigí*]" money for English classes because that would have meant less money for her husband's dream or her son's education. In weighing priorities in the family, she put herself in last place; by the use of the word "demand" she also indicated that her husband was the one who made financial decisions for the household. However, she quickly went on to depict her husband turning this sense of incompatibility upside-down, when he dramatically pointed out her *duty* as a mother to learn English in order to protect her son from assimilation into the negative aspects of U.S. culture, a life of "speaking English and making drug deals." Again, she used affective language to illustrate her recognition of her husband's moral authority over her, saying that she looked at her son and felt "a lump of tears in my throat [*un nudo en la garganta de llorar*]."

Caught between gendered constraints and gendered obligations, Camila reports seeing no choice but to creatively draw upon the media resources in English available to her. After a decade in the United States, when Camila finally found an ESL program structured around the needs of homemakers, she was already fairly competent in English. Finding the CEC a safe place in a world that still seemed dangerous, she was hesitant to move on. In fact, when in August of 2002, her CEC teacher tried to refer Camila to a more advanced ESL class in another program, Camila refused, stating a desire to be near her children's school in case of emergency (Fieldnotes, 8/15/02). Nearly 6 months later, her teacher reiterated to me, "I'm trying to get Camila to move up to a higher class, or to [community college], because she's too high for this level [intermediate]. But she won't do it" (Interview, 1/29/03).

In contrast to Camila, Trini attained some competence in spoken English while working outside the home. Certainly she was fortunate in having friendly Asian immigrant coworkers, but if she had accepted her husband's strictures on her behavior, she would never have met them. However, like many immigrant workers, Trini had no opportunity to develop English literacy on the job. Until she had children, she was content with the language skills she had. More recently, however, she had begun to think about how she could help and encourage her children to get a good education. With her children in school, she decided it was time for her to learn English, so they would see that she too was "making a little more effort" and be inspired to study. In the realm of education, as in her earlier work narratives, hard work (agentive effort) is the key to "getting ahead [*salir adelante*]." As with Camila, her investment in her children's future led her to invest in her own L2 learning (Norton Peirce, 1995).

In the narratives told by these immigrant adults, and others I talked to in the course of my research, language learning could not be separated from the gendered responsibilities of their daily lives, from their personal histories of education, work, and family life. In these narratives, immigrants invested (Norton Peirce, 1995) in L2 learning when it was congruent with other investments they had made. These investments were often strongly connected to family roles and gendered identities, such as motherhood. In order to plan appropriate curricula and classroom activities in adult ESL programs, educators need to pay close attention to students' investments. The narratives they recount are a significant resource in this endeavor.

PEDAGOGICAL IMPLICATIONS OF GENDERED LIFE HISTORY NARRATIVES

Advocating a "sociocontextual approach" to ESL instruction, Elsa Auerbach (1989) writes, "How can we draw on [learners'] knowledge and experience to inform instruction?" (p. 177). Citing Brazilian educator Paulo Freire, Auerbach points out that "literacy is meaningful to students to the extent that it relates to daily realities and helps them to act on them" (p. 166). For Freire (1999), this process takes place in dialogue between learners and educators, an encounter in which "there are neither utter ignoramuses nor true sages; there are only people who are attempting together to learn more than they now know" (p. 71). Language and literacy learning becomes relevant to adult learners when they can draw connections between the classroom and their concerns and goals outside the classroom; teachers can facilitate these connections only when they take time to listen to these concerns.

My work, along with that of Pavlenko (2001), shows the value of narrative in understanding learners' perspectives on their own trajectories of learning. Across many cultures, narrative is a resource that people draw upon to make sense of experiences. Therefore, a close examination of learner narratives can help educators to better understand tellers' perspectives on a variety of complex social and educational issues that have touched their lives over the years. Participants in my research told me multiple stories about work, love, marriage, parenting, war, politics, discrimination, poverty, shopping, reading, television, home-buying, illness, schooling, cooking, friendship; they revealed strong opinions and rich "funds of knowledge" (Moll, 1994). In many of these stories, gender played a key role. Multicultural educators can build on what learners already know (Nieto, 2002)—but only if they take the time to pay attention to where learners have come from and to where they see themselves heading.

In discussing narratives from a Salvadoran and a Mexican woman who were learners of English in 2002, it is not my intent to generalize their experiences to all L2 learners or all Latina immigrants. Instead, what I would emphasize is the value for educators of learning from their own students, who are each likely to have individual perspectives on life, learning, and identity. In order to begin to develop curricula sensitive to diverse student needs, teachers need to start by listening to their students (Kincheloe, 2003). However, listening is not always straightforward and unproblematic; nor is knowing what to do with what you have heard.

Autobiographical practices are common in ESL classes. Teachers frequently ask students to write or tell part of their life stories. At beginning levels of ESL, the Language Experience Approach involves students dictating stories to the teacher, while at more advanced levels, class newsletters collect student narratives for wider sharing. Students are frequently asked to share stories of life experience as part of cooperative learning activities. Issues raised in narratives then ideally become foci for further discussions or writing projects, though this is less common. All of these practices represent valuable learning opportunities for both students and educators, as long as teachers keep in mind that recounting autobiographical narratives can be painful for some learners and that it is wise to build in ways to opt out of assignments and discussions.

Moreover, teachers need to remember that students' classroom autobiographies are necessarily selective, partial, and shaped by perceptions of appropriateness in classroom work (Vollmer, 2000). When struggling to express themselves in a second language, students may simplify their stories, leaving out hard-to-explain complexities. One day at the CEC, the students had been assigned to write about their "favorite relative." During break time, Camila showed me a brief composition in English about her uncle and then began talking to me in Spanish about the parts of the story she had omitted, contrasting her uncle's kindness to her aunt's scolding and her father's neglect. Following a similar conversation later the same day, another student told me that they did not write the sad stories (Fieldnotes, 8/13/02). The stories I audiotaped in Spanish were far longer, more detailed, more complex, and often "sad" but still do not represent any kind of complete truth about the tellers. Therefore, teachers need to be cautious in assuming that they have understood the reality of students' lives based on student narratives; they especially need to be cautious about generalizing the narrated experience of particular learners to entire cultural groups.

Despite the fact that life-history narratives are necessarily multiple, partial, and shaped by context, I nevertheless advocate that teachers take them very seriously in developing a "reflective practice" (Schön, 1987). The experiences and perspectives in students' classroom narratives are likely to be precisely those experiences and perspectives that they want teachers to take into account in lesson and curriculum planning. By paying close attention to students' contrasting and conflicting narratives (many of which will be on gendered themes), teachers can begin to come to terms with the struggles and dilemmas which so many language learners face in their personal lives and which mediate their language learning. To quote Trini again, "So when I began to have children, I began to try to want to learn [English] so I could teach them. So then they were little and I couldn't" (Interview, 10/25/02).

As illustrated by Trini's narratives, some CEC students were beginning to redefine their positions as women within their families and communities, and this was having a positive effect on their language learning. Describing a Canadian ESL program that also served immigrant women, Harper, Peirce, and Burnaby (1996) write that "there are moments already occurring ... where issues of identity and power are problematized and not simply accommodated" (p. 17). Teachers can do their students a service by helping to create and sustain such moments, if only through asking questions like "Why? Do you agree? Does this happen to anyone else?" when students bring up potentially controversial issues. When teachers provide space for students to discuss narratives of life experience, class members can begin to examine the ways that their changing identities support or conflict with their language learning. In this way, students can learn from each other and perhaps be inspired to implement small changes in their own lives based on their classmates' examples. In addition, by asking learners to expand upon their stories, teachers can give them the space to attempt to express in their second language some of the narrative complexities that they may have initially edited out. In this way, teachers can promote L2 development by creating opportunities for learners to draw upon the full range of their L2 grammatical and lexical resources as they negotiate meaning with their classmates (Gass, Mackey, & Pica, 1998; van Lier, 2000).

Recognizing that language learning is often mediated by social positioning, ESL teachers and programs need to accept their responsibility to serve learners who are constrained by gender ideologies as well as those who are beginning to question them (often the same people!). To give one example, many immigrant women will not be able to attend classes unless child care is provided. However, even as teachers help students to work within gender constraints, they need not take traditional gender ideologies as given and unquestionable. As the narratives in this article illustrate, even within one immigrant "community," ideologies vary between families and individuals; they are subject to change over time. The narratives that learners tell in the classroom about their lives are potentially a contribution to a dialogue in which teachers and students work together to "name the world" in order to transform it (Freire, 1999, p. 68). Through dialogue, the interconnections between classroom practice, social positioning, and students' lived experience can become a resource for language learning rather than a constraint to be overcome.

ACKNOWLEDGMENTS

I thank the participants in this research, especially "Camila" and "Trini." I also appreciate my transcriber, Luis Solano; my friends Deborah Palmer, Paige Ware, and Jessica Zacher, who read early drafts of this article; and guest editor Juliet Langman and two anonymous reviewers for comments that helped shape my revisions.

ENDNOTES

[1]All names are pseudonyms.

[2]Quotes from participants are translated from Spanish by the author. The Spanish text is omitted from this article due to space limitations, except for key phrases.

REFERENCES

Alvarez, J. (1998). *Something to declare.* Chapel Hill, NC: Algonquin Books.

Auerbach, E. (1989). Toward a social-contextual approach to family literacy. *Harvard Educational Review, 59,* 165–181.

Bogdan, R. C., & Biklen, S. K. (1998). *Qualitative research for education: An introduction to theory and methods* (3rd ed.). Boston: Allyn & Bacon.

Boyatzis, R. E. (1998). *Transforming qualitative information: Thematic analysis and code development.* Thousand Oaks, CA: Sage.

Bruner, J. S. (1990). *Acts of meaning.* Cambridge, MA: Harvard University Press.

Bucholtz, M., Liang, A. C., & Sutton, L. (Eds.). (1999). *Reinventing identities: The gendered self in discourse.* New York: Oxford University Press.

Cortazzi, M. (1993). *Narrative analysis.* London: Falmer.

Cumming, A., & Gill, J. (1992). Motivation or accessibility? Factors permitting Indo-Canadian women to pursue ESL literacy instruction. In B. Burnaby & A. Cumming (Eds.), *Sociopolitical aspects of ESL education in Canada* (pp. 241–252). Toronto: OISE Press.

Eckert, P., & McConnell-Ginet, S. (1992). Think practically and look locally: Language and gender as community-based practice. *Annual Review of Anthropology, 21,* 461–490.

Ehrlich, S. (1997). Gender as social practice: Implications for SLA. *Studies in Second Language Acquisition, 19,* 421–446.

Fairclough, N. (1992). *Discourse and social change.* Cambridge, England: Polity.

Freire, P. (1999). *Pedagogy of the oppressed* (20th Anniversary ed.). (M. B. Ramos, Trans.). New York: Continuum Publishing.

Gass, S., Mackey, A., & Pica, T. (1998). The role of input and interaction in second language acquisition. *Modern Language Journal, 82,* 299–307.

Goldstein, T. (1997). *Two languages at work: Bilingual life on the production floor.* Berlin: Mouton.

Harper, H., Peirce, B., & Burnaby, B. (1996). English-in-the-workplace for garment workers: A feminist project? *Gender and Education, 8,* 5–19.

Hoffman, E. (1989). *Lost in translation: A life in a new language.* London: Penguin.

Kibria, N. (1993). *Family tightrope: The changing lives of Vietnamese Americans.* Princeton, NJ: Princeton University Press.

Kincheloe, J. L. (2003). *Teachers as researchers: Qualitative inquiry as a path to empowerment* (2nd ed.). London: Routledge.

Labov, W. (1972). The transformation of experience in narrative syntax. In W. Labov (Ed.), *Language in the inner city: Studies in the black vernacular* (pp. 354–397). Philadelphia: University of Pennsylvania Press.

Martin, J. R. (2000). Beyond exchange: Appraisal systems in English. In S. Hunston & G. Thompson (Eds.), *Evaluation in text: Authorial stance and the construction of discourse* (pp. 142–175). New York: Oxford University Press.

Moll, L. (1994). Mediating knowledge between homes and classrooms. In D. Keller-Cohen (Ed.), *Literacy: Interdisciplinary conversations* (pp. 385–410). Cresskill, NJ: Hampton.

Nieto, S. (2002). *Language, culture, and teaching: Critical perspectives for a new century.* Mahwah, NJ: Lawrence Erlbaum Associates, Inc.

Norton Peirce, B. (1995). Social identity, investment and language learning. *TESOL Quarterly, 29,* 9–31.

Ochs, E., & Capps, L. (1996). Narrating the self. *Annual Review of Anthropology, 25,* 19–43.

Pavlenko, A. (2001). Language learning memoirs as a gendered genre. *Applied Linguistics, 22,* 213–240.

Rockhill, K. (1993). Gender, language and the politics of literacy. In B. Street (Ed.), *Cross-cultural approaches to literacy* (pp. 156–175). Cambridge, England: Cambridge University Press.

Schecter, S. R., & Bayley, R. (2002). *Language as cultural practice: Mexicanos en el Norte.* Mahwah, NJ: Lawrence Erlbaum Associates, Inc.

Schön, D. A. (1987). *Educating the reflective practitioner.* San Francisco: Jossey-Bass.

Scribner, S. (1997). *Mind and social practice: Selected writings.* Cambridge, England: Cambridge University Press.

van Lier, L. (2000). From input to affordance: Social-interactive learning from an ecological perspective. In J. Lantolf (Ed.), *Sociocultural theory and language learning* (pp. 245–259). New York: Oxford University Press.

Vollmer, C. G. (2000). *Classroom contexts for academic literacy: The intersection of language and writing development in secondary ESL classrooms.* Unpublished doctoral dissertation, University of California, Berkeley.

Wenger, E. (1998). *Communities of practice: Learning, meaning, and identity.* Cambridge, England: Cambridge University Press.

JOURNAL OF LANGUAGE, IDENTITY, AND EDUCATION, *3*(4), 313–322

BOOK REVIEWS

Bilingual Couples Talk: The Discursive Construction of Hybridity. Ingrid Piller, Amsterdam: Benjamins, 2002, xii + 314 pages, $90.00 (hardcover).

Marya Teutsch-Dwyer
St. Cloud State University

This highly readable and carefully researched book on bilingual couples' talk is a must-read for all interested in issues of interaction between language and identity, the significance of belief systems for language use, and the polyphony of discourse, that is, how interlocutors position themselves vis-à-vis larger discourses. Aligning her approach with poststructuralism and social constructivism, and drawing upon the Bakhtinian idea that conversational meaning cannot be understood without reference to larger discourses, Ingrid Piller examines the ways in which bilingual couples perceive and perform their identities in different situations, what beliefs enable or constrain their language choices and bilingualism, and how conflicting ideologies are played out in private conversations. The book highlights the constant and powerful contradictions that underlie the relationship between language choice and personal relationships, between language use and emotional connection, and between private and public discourses. While language choice among bilingual couples has been previously investigated (e.g., Pauwels, 1985), this study is unique and valuable in that its research context differs from the societal contexts in which language maintenance and language shift have normally been studied. In contrast to the other studies, this research does not concern itself with intergenerational transmission. Neither of the two languages—German and English—in this study can be labeled a "minority language," and neither language has historically been in a mutually dominated/dominant relationship. In addition, the participants' Whiteness, class position, educational level, and language status do not conform to the typical parameters of the populations in studies on language maintenance and shift. This thorough book consists of 10 chapters, notes on transcription, and notes on selected chapters. It is divided into two broad sections: the first one (chapters 2–4) deals with data collection and introduces the participants; the second one (chapters 5–9) describes the findings of the research.

In chapter 1, "Researching Bilingual Couple Talk: A Discourse-Analytic Approach to Language Contact," Piller explains what prompted her to do research on bilingual couple talk (rather than on individuals' talk in bilingual relationships). While marriage used to be regarded as an economic unit in which to raise a family, today's relationships are more likely to seek romance, friendship, and common interests. Communication has, therefore, become a central and constitutive factor in the make-up of modern relationships. The study of bilingual couple talk is no less important, as the numbers of cross-cultural relationships is on the rise worldwide. The chapter outlines the author's theoretical framework, which, among others, includes a crucial premise that "identity mediates language use, and ideology mediates identity" (p. 13).

Chapter 2, "What We Know: Bilingual Couples in Linguistic Research," provides the readers with definitions of the main terms, such as "majority language," "minority language," "language maintenance" and "language shift," used throughout the book. Piller explores how national identities, native and nonnative speaker status, and ideologies of cross-cultural couplehood are played out in actual conversations. The author points out how terms such as "intermarriage" and "exogamy" are negatively collocated with concepts of challenge and risk, or "threat to survival," to name just a few. In examining the various outcomes of language contact and language shift in the English and German bilingual couples, the author argues that the impact of the majority language on a couple's language choice is by no means as straightforward as the research so far may have implied; it is complex and mediated by different factors. Proficiency, for one, plays a crucial role. Many partners are already bilingual when they enter the relationship; the choice of language may then depend on other factors, such as desired identities, gender identities, and language ideologies. The impact of the macrosocietal context in which couples make language decisions on a microlevel may also play a decisive role.

Chapter 3, "'It Needs to be Natural': Building a Corpus," gives the reader an interesting glimpse into the data collection details and the numerous, often unexpected, problems along the way. For example, despite the fact that the couples were given instructions to record themselves in as natural circumstances as possible some returned empty tapes, one with a note saying they "could not find an occasion to make a recording of any kind." Other recordings were far from "natural." Due to these and other similarly problematic data, Piller changed her original method and, instead, employed what she calls "a discussion paper" with a set of interview questions, which were to elicit information about the couples' linguistic practices and, concurrently, samples of actual talk. The final corpus consisted of 36 core couples' talk, with a total length of 18 hours and 44 minutes, all transcribed in full. Supplementary data consisted of an additional 2 hours and 7 minutes of spoken interaction, numerous written texts, which included letters and e-mails written to the author, postings to *biling-fam*, 13 issues of *The Bilingual Family Newsletter*, 9 issues of *Polyglot*, 4 issues of *Currents*, 5 issues of *In Touch*, 20 issues of *The Written Word*, plus Web sites, newspaper articles, and books.

Chapter 4, "The Couples," provides social biographical information about the couples in the study. The information comes from the couples' responses to the questionnaires, from their recorded conversations, and other communication Piller had with them during the project. "In keeping with the poststructuralist framework embraced in this work" (p. 59), the researcher created rich social portraits of 36 couples. Histories of language learning and language use, (perceived) native languages, the couples' reasons for choosing one language over another over time and in different situations, their educational levels and current professions, family and/or religious backgrounds were among the prominent and consistent features discussed.

In chapter 5, "'I Speak English Very Well': Linguistic Backgrounds," Piller scrutinizes the participants' language proficiencies and language ideologies. She states that she approached the analysis from four different perspectives: (a) beliefs about the native language, (b) trajectories and ideologies of second language (L2) learning, (c) discourses about success in L2 learning, and (d) diglossia as the macrolinguistic context in which choices are made. Set within these parameters, the author shows how complex and unclear to the participants themselves their own linguistic situations are. For example, one third of all the participants claimed a different first language (L1) on the questionnaire than during the "discussion paper" on the tape. These differences are a result of ideologies of symbolic language ownership, with writing abilities being the strongest factor in this perception of ownership. Another interesting finding is that of the participants' perceptions about the best way to learn a language. Most of the German-speaking participants, who were exposed to many years of intensive studies of English in a formal environment, claimed that the only way to learn a language is through naturalistic encounters. However, unlike the naturalistic learner of English in my study (Teutsch-Dwyer, 2001), who also strongly believed in acquiring a language in a naturalistic environment, the L1 German speakers of English not only achieved high levels of ultimate proficiency but also displayed a relatively strong sense of L2 ownership. This chapter also supports the author's argument that the participants' involvement with their L2 preceded (rather than followed) their cross-cultural relationship.

Chapter 6, "'We Speak Bilingually': Language Choice," successfully provides counterexamples to a pervasive hypothesis which states that "intermarriage leads to language shift." Piller reminds us that the simplistic approach to the question of language choice in bilingual couples, which assumes that the majority language would be the "natural" choice for bilingual couples, is simply erroneous. As her data show, most of the couples do not choose the majority language, nor do they make a conscious decision which language to choose. Language choice (the majority language, the minority language, or the mixed code) results from more complex reasons. These include, for example, habit (i.e., continuing to use the language of their first interaction, in which case English is a more likely language to be used, whether the couple met in a German or English-speaking country), compensation (i.e., using one's native language in return for giving up one's own homeland), or identity (preserving a desired self-image through the use of one of the languages). Hybridity, or language mixing, encompasses broader and more complex linguistic practices. The couples have varied evaluations of language mixing, they often seem to be unclear about how it is done and in what situations, and as a result, their statements are frequently contradictory.

Chapter 7, "'We are Citizens of the World': Identity and Cross-Cultural Couplehood," describes how the bilingual couples view their cultural, national, and linguistic identities, both in public and private discourses. Piller argues that while the couples value highly the performance of a common "couple identity," their discourse of national identity, in contrast, tends to create difference and a threat to their couple identity. To counterbalance this difference, which is a consequence of the fact that the partners are members of different national groups, the couples make an effort to focus on other nonnational identities, such as age, class, education, or profession, to create similarities. Through employing different strategies, such as constructing similarity and deconstructing difference (for example, by presenting it as desirable), the couples' ultimate goal seems to be to perform their "couple identity."

In chapter 8, "'The Talk is Essential': Doing Couplehood," Piller continues her discussion of the performance of couplehood, which is enacted through the couples' conversational style. According to her findings, on a metalinguistic level the couples consider their private language a constitutive factor in creating their relationship. The conversational data demonstrate that couplehood is performed on the collaborative floor; couples use several strategies, such as filling in the missing words (similar to scaffolding), repetition, offering or asking for clarification after a misunderstanding, and the joint telling of language narratives. It does not mean collaboration is ever-present. Moments of discord often result from a situation in which one partner turns a collaborative floor into a single floor. These "conversational troubles," however, tend to occur only as a result of discord over nonconversational issues.

Chapter 9, "'The Doors of Europe Will be Open to Them': Private Language Planning," focuses on the couples' language planning for their children and their linguistic future. It is clear that the participants are not only familiar with research on childhood bilingualism (for example, they are familiar with the one-parent one-language principle) but they also show high levels of commitment to their children's bilingual education. The author welcomes this positive attitude towards bilingualism in her data; however, she cautions against extreme ideologies, which, when positive, may bring about unfortunate feelings of parental failure when their children reject bilingualism or turn out to be less than perfectly balanced bilinguals. The concluding chapter, "'I'm a Hybrid': Hybrid Identities, Multiple Discourses," provides summaries and ties together the major threads of this study.

This original and thought-provoking work will be of great interest to readers in several disciplines, particularly sociolinguists, foreign language, second language, and bilingualism scholars, psycholinguists, and psychologists. The book's novel findings and systematic research methodology will no doubt spur further research.

It can be used in graduate and undergraduate courses. Finally, *Bilingual Couples Talk* will make a great reading for anyone who is in a bilingual relationship or considers one in the future.

REFERENCES

Pauwels, A. (1985). The effect of exogamy on language maintenance in the Dutch-speaking community in Australia. *ITL Review of Applied Linguistics, 66,* 1–24.
Teutsch-Dwyer, M. (2001). (Re)constructing masculinity in a new linguistic reality. In A. Pavlenko, A. Blackledge, I. Piller, & M. Teutsch-Dwyer (Eds.), *Multilingualism, second language learning and gender* (pp. 175–198). Berlin: Mouton.

The Verbal Communication of Emotions: Interdisciplinary Perspectives. Susan Fussell (Ed.), Mahwah, NJ: Lawrence Erlbaum Associates, Inc., 2002, 294 pages, $29.95 (softcover).

Aneta Pavlenko
Temple University

In the past few years there has been a veritable explosion of edited volumes and monographs on the relationship between language and emotions (Athanasiadou & Tabakowska, 1998; Harkins & Wierzbicka, 2001; Kövecses, 2000; Palmer & Occhi, 1999; Wierzbicka, 1999). While emotions have been studied in mainstream psychology for at least a few decades, it was not until recently that the role of language in construction of emotions came to the foreground in this area, due to contributions from linguistic anthropology, cognitive linguistics, and cultural and discursive psychology. Drawing on rich data from a variety of languages and contexts, recent scholarship highlights the role of conceptual metaphors and cultural models in construction of emotions, pointing to similarities and differences in emotion vocabularies and repertoires across languages and cultures. In view of such an abundance of scholarly work, it would appear that one more edited volume could hardly make a difference. And yet Susan Fussell, the editor of *The Verbal Communication of Emotions,* has managed to produce a very important collection that familiarizes the readers with major advances in this area of research. Rather than adding a few more studies and reflections to the general body of knowledge, the contributors to the volume—many of them key players in the field—take stock of what has been accomplished to date and offer programmatic suggestions that aim to push the field forward.

The book opens up with a chapter by the editor, Susan Fussell, "The Verbal Communication of Emotion: Introduction and Overview." The chapter offers a rationale for the focus on verbal expression of emotions, a discussion of disciplines and approaches where such research takes place, and an overview of the chapters. The rest of the volume is divided into three sections, the first of which deals with background theory, the second with figurative language use, and the third with social and cultural aspects of emotional communication.

Part 1, "Theoretical Foundations," contains three papers. In "Explicating Emotions Across Languages and Cultures: A Semantic Approach," Cliff Goddard describes an integrated approach to the study of the language of emotions, which combines semantic and pragmatic inquiry. He starts with an excellent argument against the dangers of an ethnocentric approach, which takes as a starting point English emotion words and folk categories. Then he carefully reviews common experimental techniques and survey instruments, pointing to multiple ways in which the use of "translation equivalents" leads to creation of flawed test instruments and compromises the validity of experimental results. To remedy the problem of "loss in translation," Goddard suggests the use of the "natural semantic metalanguage" (NSM) framework, developed by Wierzbicka and applied extensively to emotion semantics by Goddard, Wierzbicka (1999) and associates (cf. Harkins & Wierzbicka, 2001). In this framework, the meaning of a word is paraphrased in a metalanguage, which currently consists of about 65 "semantic primes," that is, words that are found in all or most of the world's languages. To illustrate the use of the NSM framework, Goddard brings up a variety of examples from English, Malay, Polish, and Japanese. Subsequently, he makes another important argument, namely, that to fully understand the language of emotions, we need to look not only at lexical semantics but also at cultural pragmatics. In his view, this can be done best through a "cultural scripts" approach, also developed by Wierzbicka. To illustrate the approach, Goddard uses American, Polish, Malay, and Japanese cultural scripts.

In the next chapter, "Integrating Verbal and Non-Verbal Emotion(al) Messages," Sally Planalp and Karen Knie take a look at another theoretical and methodological challenge—the interaction between verbal and non-verbal cues in communication of emotions. The authors review three paradigms that offer somewhat distinct views on whether and how the two sets of cues may be integrated. The expressive paradigm conceptualizes emotion as a substance that "leaks out" through an array of verbal and non-verbal manifestations and thus posits stable links between emotions and co-occurring cues (e.g., joy, positive exclamations, and smiles). The conventional paradigm views emotion as the content of a message sent through a combination of verbal and non-verbal cues. The rhetorical paradigm sees emotional messages as phenomena that evolve in social context and rely on verbal and non-verbal cues to achieve social goals. The authors carefully review the advantages and disadvantages of each model and conclude that the most productive research will be conducted at the intersection of the three approaches, where expressivity, communication, and social context can all be taken into consideration in order to explain how and where verbal and non-verbal emotion cues cohere or conflict.

The third chapter in this section, "How to Do Emotions with Words: Emotionality in Conversations," by Reinhard Fiehler, outlines an interactional approach to the study of affective communication. The chapter starts out with a critique of linguistic theories that have trouble in accommodating emotionality and of methodological approaches that cannot easily accommodate the study of everyday conversation. Then, Fiehler offers a coherent framework that posits four types of rules that regulate the occurrence and manifestation of emotions, and two key communicative strategies, expression and thematization of emotions. What is particularly useful about the chapter is that the theoretical framework is subsequently translated into a six-step analytical approach described in detail and illustrated in a sample analysis of a therapy conversation. It was a pleasant surprise to see that the chapter is a translation from German, made specifically to render the author's work more accessible to English-speaking researchers. This editorial strategy is definitely worthy of emulation, as it introduces important work conducted in languages other than English and opens up the pages of English-language publications to scholars who do not or choose not to write in English.

Part 2, "Figurative Language in Emotional Communication," contains four chapters. The first, "Emotion Concepts: Social Constructionism and Cognitive Linguistics," is written by Zoltán Kövecses, an internationally known expert in the field. The focus of the chapter is on similarities and differences between two approaches to the language of emotions—that is, discursive psychology, advocated by Rom Harré, and cognitive linguistics, espoused by the author. Acknowledging some similarities between the two approaches, Kövecses then points to the weaknesses of discursive psychology, or social constructionism. Among the key weaknesses, in his view, are the focus on emotion words and categories which obscures the contribution of metaphoric language and a privileging of linguistic and cultural relativity at the expense of universality. Kövecses argues that a large part of emotional conceptualization is universal and grounded in embodied experiences, and he uses metaphoric expressions from English, Chinese, Hungarian, and Wolof to prove his point.

The next chapter, "What's So Special about Figurative Language in Emotional Communication?" by Raymond Gibbs, John Leggitt, and Elizabeth Turner, reminds the reader that emotions are constructed not in the heads of imaginary speakers but in conversational interaction. The authors review psycholinguistic studies that examine how people use figurative language in talking about their own and others' emotional states and experiences and how they interpret metaphorical expressions and react to them. The results of this inquiry suggest that speakers use figurative language strategically to express nuances of emotion states and elicit specific emotional reactions. Of particular interest to the authors are the uses of irony. Their review places irony squarely within the study of the language of emotions, showing that speakers generally choose ironic statements because of how they affect addressees' emotions.

It is at this point that the focus of the book shifts from more general to more specific, from theoretical and methodological issues in the study of verbal communication of emotions to communication in therapeutic contexts. Chapter 7, "Conflict, Coherence, and Change in Brief Psychotherapy: A Metaphor Theme Analysis," by Lynne Angus and Yifaht Korman, is a study of metaphors in communication between two therapists and their two clients who had shown a high improvement rate. The authors argue that the change in metaphors used by the clients is reflective of the change in their experiencing and conceptualization of themselves and others. The theme is continued in chapter 8, "Conventional Metaphors for Depression," by Linda McMullen and John Conway. Rather than examining all of the uses of metaphor, these researchers focus on the figurative language used by the clients to describe their experience of depression. The authors argue that the dominant metaphor identified in their corpus, DEPRESSION IS DESCENT, serves to position the speakers in opposition to what is culturally valued in the Western society.

Part 3, "Social and Cultural Dimensions," is comprised of four chapters. Chapter 9, "Emotion, Verbal Expression, and the Social Sharing of Emotion," by Bernard Rimé, Susanna Corsini, and Gwénola Herbette, offers a lively and comprehensive discussion of research on communication that takes place after a particular emotional episode. Based on evidence from interview, diary, and experimental data, the authors argue that, upon participation in an emotional event, people tend to share their emotional experiences with others. These findings hold across a range of ages and cultures, with some cross-cultural differences observed in sharing modalities. The next chapter, by Jeffrey Pittam and Cynthia Gallois, is "The Language of Fear: The Communication of Intergroup Attitudes in Conversations about HIV and AIDS." The authors focus their attention on one topic in affective communication, the language used to express fear of AIDS. Their empirical study of discussion groups set up among Australian university students shows that speakers' lexical choices reflect not only their feelings about AIDS but also their social, gender, national, and sexual identities. The authors of chapter 11, "Rewards and Risks of Exploring Negative Emotion: An Assimilation Model Account," are Lara Honos-Webb, Linda Endres, Ayesha Shaikh, Elizabeth Harrick, James Lani, Lynne Knobloch-Fedders, Michael Surko, and William Stiles. The chapter familiarizes the reader with the Assimilation Model framework which outlines a systematic series of changes in a client's representations of problematic experiences. Next, they present two case studies that show how the framework can be used to track clients' progress. The results of these studies suggest that expression of negative emotions in the context of psychotherapy can lead to symptom reduction. The authors also point to studies that suggest that in some cases expression of negative emotion can lead to disruptive outcomes. Chapter 12, "Blocking Emotions: The Face of Resistance," by Kathleen Ferrara, is an appropriate final chapter. While the rest of the chapters focus on emotional expression, Ferrara examines how clients resist therapeutic suggestions, such as "let yourself cry," and, in fact, hold emotions back. Using a discourse analytic approach, she shows how one can identify and operationalize resistance in conversation and offers suggestions for more effective therapeutic communication.

All in all, this interdisciplinary volume with contributions from all over the world could be seen as a primer to the field that introduces novices to cutting-edge theories and up-to-date findings and raises new challenges for those who are already involved in research on affective communication. At the same time, the fact that this book review is aimed at the *Journal of Language, Identity, and Education* audience made me rather uncomfortable, since educational contexts are, in fact, neglected by the book's editor and the volume contributors. Rather, the secondary focus of the volume is on therapeutic contexts and clinical interaction. Undoubtedly, health care communication should be of major concern for scholars and practitioners. Yet it would also be desirable to extend the focus of attention to communication of emotions in the workplace and in educational contexts.

Similarly, while the cross-linguistic focus of this volume is commendable (especially in view of the earlier lack of sensitivity to cross-linguistic differences), one cannot help but notice that the research on language and emotions is continuously conducted from a monolingual perspective, namely, with the focus on monolingual speakers of particular languages. This of course is not surprising: Western scholars in social sciences and humanities, in particular in linguistics and psychology, have been traditionally apprehensive about working with bi- and multilingual participants and informants, concerned that their perceptions, intuitions, and performances may exhibit "impure" knowledge or "incomplete competence" in the language in question and convinced that languages are better studied in an "idealized" case of monolingual competence. At times, they have also been guided by the assumption that whatever applies to monolinguals should be applicable to speakers of more than one language. At other times, the guiding assumption was that one needs to first figure out how things work out in monolingual contexts, and only then can we take a look at the messy language contact situations. In either case, monolingualism was positioned as "the norm" and bi- and multilingualism as an exception, best examined on the margins of the mainstream scholarship.

In view of rapidly increasing linguistic diversity in the global marketplace, such assumptions are no longer simply misguided. They are downright harmful, as they disenfranchise those who need attention most: nonnative speakers, speakers of minority languages and nonstandard varieties, whose communication patterns may deviate from the White middle-class mainstream "norm" in the classroom, in the workplace, in clinical interaction, or in the legal system. They are also conceptually flawed, as there is no guarantee that when and if we figure out "how it all works," theories developed in monolingual contexts would apply to multilingual speakers. Rather, any investigation of linguistic phenomena, including that of the language of emotions, needs to be centrally concerned with the way the phenomena in question play out in multilingual and heteroglossic contexts.

I recommend Fussell's volume to the *Journal of Language, Identity, and Education* readers as an excellent guide to up-to-date research on communication of emotions and as an inspiration for a new wave of interdisciplinary scholarship, which will explore how bi- and multilinguals communicate emotions in public and private spaces, how we can teach and learn new emotion repertoires, how emotion concepts may be transformed in second language socialization (Pavlenko, 2002), how bilingual clients behave in therapy (Santiago-Rivera & Altarriba, 2002), and how linguistically diverse students may express emotions in the classroom (Harkins, 1990).

REFERENCES

Athanasiadou, A., & Tabakowska, E. (Eds.). (1998). *Speaking of emotions: Conceptualisation and expression.* The Hague, Netherlands: Mouton.

Harkins, J. (1990). Shame and shyness in the Aboriginal classroom: A case for "practical semantics." *Australian Journal of Linguistics, 10,* 293–306.

Harkins, J., & Wierzbicka, A. (2001). *Emotions in crosslinguistic perspective.* Berlin: Mouton de Gruyter.

Kövecses, Z. (2000). *Metaphor and emotion: Language, culture, and body in human feeling.* Cambridge, England: Cambridge University Press.

Palmer, G., & Occhi, D. (1999). *Languages of sentiment: Cultural constructions of emotional substrates.* Amsterdam: Benjamins.

Pavlenko, A. (2002) Bilingualism and emotions. *Multilingua, 21,* 45–78.

Santiago-Rivera, A., & Altarriba, J. (2002). The role of language in therapy with the Spanish-English bilingual client. *Professional Psychology: Research and Practice, 33,* 30–38.

Wierzbicka, A. (1999). *Emotions across languages and cultures: Diversity and universals.* Cambridge, England: Cambridge University Press.

JOURNAL OF LANGUAGE, IDENTITY, AND EDUCATION, 3(4), 323
Copyright © 2004, Lawrence Erlbaum Associates, Inc.

ACKNOWLEDGMENTS

The editors of the *Journal of Language, Identity, and Education* thank our two editorial assistants, Kelly L. Graham (University of Texas, San Antonio) and Wayne E. Wright (Arizona State University), who have played an important role in producing Volume 3 of the *Journal of Language, Identity, and Education*. We also acknowledge the work of our production editor, Janet Roy of Lawrence Erlbaum Associates, Inc., and the many other professionals at Lawrence Erlbaum who have played an important role in the production of the journal. Finally, we are happy to acknowledge the contribution of the following scholars (including members of the Editorial Advisory Board, whose names are preceded by asterisks) who reviewed manuscripts in the period ending December 2003.

Muhammad Amara
*Dwight Atkinson
*Colin Baker
*John Baugh
*Robert Bayley
Jill Bell
*Sarah Benesch
Felecia Briscoe
*Janina Brutt-Griffler
*Suresh Canagarajah
*Ursula Casanova
John Flowerdew
*James Paul Gee
*Nancy H. Hornberger
Thom Huebner
Yasuko Kanno
*Kimi Kondo-Brown

Ryuko Kubota
*Juliet Langman
Eric Margolis
*Brian Morgan
*Bonny Norton
Aneta Pavlenko
*Alastair Pennycook
Ingrid Piller
Vai Ramanathan
*Stanley Ridge
Kevin Rocap
Keiko K. Samimy
*Otto Santa Ana
Elana Shohamy
Mary Lee Smith
*Kamal F. Sridhar
*Guadalupe Valdés

Contributor Information

Editorial Scope: The *Journal of Language, Identity, and Education* (*JLIE*) is an international forum for interdisciplinary research that is grounded in theory and of interest to scholars and policymakers. Education plays a central role in promoting social development, stability, integration, and equity in a linguistically and culturally diverse world. Policy decisions in educational settings today often require an understanding of the relations between home language/variety and school language/variety, relations of language, ethnicity, and gender identity construction, societal attitudes toward languages/varieties, and differential performance across groups. *JLIE* seeks cutting edge interdisciplinary research from around the world, reflecting diverse theoretical and methodological frameworks and topical areas and solicits articles that deal with the following topics:

- Educational policies and approaches that explicitly address various dimensions of diversity;
- The formation and consequences of identities in educational and other social contexts;
- Language policies and linguistic rights in educational contexts;
- The role of indigenous languages/varieties in education;
- Critical studies of literacy policies, national literacy and biliteracy demographics, the socioeconomic and political significance of literacy, and societal expectations regarding literacy;
- Research on the relation between home/local linguistic and cultural socialization and schooling;
- Critical and comparative analyses of official and legal frameworks for educational policies and practices in diverse settings;
- Critical studies of school and community attitudes and expectations about schooling;
- Critical studies about bias in schooling practices;
- Research on educational practices that promote educational equity for diverse student populations;
- The role of ideologies in educational language and cultural policies; and
- Group-specific studies on special needs/issues and on effective policies and practices.

Submission Guidelines: Manuscripts should be double-spaced—including title page, text, tables, charts, references, notes, and appendixes—and must adhere to the guidelines of the *Publication Manual of the American Psychological Association* (5th ed.). The first page should include the title, name(s), and affiliation(s) of author(s) and full contact addresses for correspondence (including e-mail). The second page should include the title (but no author identification), an abstract of not more that 150 words, a list of up to 6 key words, and a word count. Use either American or British spelling consistently within an article. Manuscripts must be single-sided, typed on 8½" × 11" or A4 paper and should normally be no more than 25 to 30 double-spaced pages (including references, notes, and tables). Minimize the number of notes.

Tables and figures should be placed after the references, each on a separate page. Mark in the text where they occur. Camera-ready copy is required for publication of figures. Please contact the editors to inquire about undertaking a review; unsolicited reviews will not be accepted. Reviews should be between 1,500 and 2,000 words. Guidelines are the same as for articles. Only original work not previously published and not currently under review elsewhere will be considered. Contributions should be in English and will be reviewed anonymously.

After acceptance, a final version of the article will be required on diskette as well as in hard copy. Send five (5) copies of the manuscript to either editor:

Thomas Ricento, Division of Bicultural-Bilingual Studies, College of Education and Human Development, University of Texas, 6900 N. Loop 1604 West, San Antonio, TX 78249–0653. Fax: 210-458-5962. E-mail: tricento@utsa.edu

Terrence Wiley, Division of Educational Leadership and Policy Studies, College of Education, Arizona State University, Main Campus, P.O. Box 872411, Tempe, AZ 85287–2411. Fax: 480-965-1880. E-mail: twiley@asu.edu

Book Review Editor: Vaidehi Ramanathan, Department of Linguistics, University of California, Davis, CA 95616. Phone: 530-752-0191. E-mail: vramanathan@ucdavis.edu

Milton Keynes UK
Ingram Content Group UK Ltd.
UKHW022109141024
449569UK00031B/1847